U.S. Department of Justice
Office of Justice Programs
National Institute of Justice

I-ADAM IN EIGHT COUNTRIES

APPROACHES AND CHALLENGES

RESEARCH REPORT

U.S. Department of Justice
Office of Justice Programs
810 Seventh Street N.W.
Washington, DC 20531

John Ashcroft
Attorney General

Deborah J. Daniels
Assistant Attorney General

Sarah V. Hart
Director, National Institute of Justice

| **Office of Justice Programs** World Wide Web Site *http://www.ojp.usdoj.gov* | **National Institute of Justice** World Wide Web Site *http://www.ojp.usdoj.gov/nij* |

I–ADAM IN EIGHT COUNTRIES
Approaches and Challenges

Editor
Bruce Taylor

Contributing Authors
Vicknasingam Balasingam, Trevor Bennett, Henry H. Brownstein, Luis Caris, Willemijn Garnier, Doug Johnson, Peter Klerks, D. Locke, Antoinette Louw, Toni Makkai, Neil McKeganey, Visweswaran Navaratnam, Hilal Hj. Othman, Charles Parry, Janine Plaisier, Andreas Plüddemann, Bruce Taylor, Elna van Niekerk, and Gina Weir-Smith

May 2002
NCJ 189768

National Institute of Justice

Sarah V. Hart
Director

Points of view or opinions stated in this document are those of the authors and do not represent the official position or policies of the U.S. Department of Justice.

The National Institute of Justice is a component of the Office of Justice Programs, which also includes the Bureau of Justice Assistance, the Bureau of Justice Statistics, the Office of Juvenile Justice and Delinquency Prevention, and the Office for Victims of Crime.

FOREWORD

James O. Finckenauer

James O. Finckenauer is director of the International Center of the National Institute of Justice. He is on leave from the Rutgers University School of Criminal Justice, where he is a professor of criminal justice.

The International Arrestee Drug Abuse Monitoring (I–ADAM) program began just over 2 years ago. *I–ADAM in Eight Countries: Approaches and Challenges* is the first look at the general results of this international experiment. The compendium provides an update—a kind of status report—on the state of I–ADAM's implementation.

It also provides an opportunity to address the broader subject of international comparative research. Research purporting to examine crime and justice issues across national boundaries, while occasionally lauded, has been relatively rare. This rarity is attributable in part to the great difficulties faced by any would-be comparativists; some of these difficulties are of sufficient gravity to call into question the very worth of any such undertakings.

Skeptical questions often raised about comparative research include: What is its value? What are its contributions to our body of knowledge about crime and criminals? What are its pitfalls? And perhaps most important, are those pitfalls so potentially damaging to research credibility as to render any findings and conclusions from international comparative studies meaningless? Although this initial and modest effort by the I–ADAM program can serve as a useful case example, it will obviously not definitively answer the criticisms and questions. It will, however, permit us to begin to address a number of issues subsumed within some of the questions.

Before turning to cross-national research in general and the I–ADAM studies in particular, let us first consider the issue of the overall worth of any research information that is "comparative." Findings and results of research can be grouped into three types of information: descriptive, comparative, and explanatory. Although any research results must begin with descriptive information, mere descriptions, by themselves, tell us very little. For example, let us say we find that 30 percent of arrested drug users recidivate. What are we to make of this isolated piece of information? Is that figure high, low, or average? Is it increasing or decreasing? Without some basis for comparison, we do not know. Thus, we see the need for comparative information—literally, an answer to the question, "Compared with what?" Before/after studies, multisite studies, and experimental/control studies, for instance, all introduce a comparative frame of reference. Of the three information types, explanatory information is of the highest order in facilitating understanding. But to obtain explanatory information, one must have rigorous research designs that enable testing causal assumptions so as to actually explain the phenomenon in question. Such designs are quite expensive and difficult to maintain in criminological studies.

As the above discussion implies, comparative research can occur in many forms. Only one of these—albeit the one most thought of as "comparative research"—involves cross-national work. Cross-national comparative studies introduce cultural and national diversity into the research design. They enable us to address questions about the universality of crime and criminal behavior, the criminal justice system response, the offender and community reactions to that system response, and the universality of the interactions among these factors. With respect to drug abuse, the focus of the research reported here, we can ask about differences across countries in drugs, drug use markets and prices, or the age and gender of users. We can further ask about differences in the relationship between drug use and criminal offending, the type of crimes committed, and, again, offender age and gender. As the introduction to this compendium points out, comparing the prevalence of drug use among arrestees in different countries is valuable to help identify invariant factors that predict drug use across countries and general patterns of human behavior that are not culturally specific.

I–ADAM has considerable potential for creating a research-based policy platform. The value in having answers to questions such as those just mentioned lies in finding cross-national differences that lend themselves to policy interventions. For example, if specific criminal justice responses or prevention strategies are shown to be effective in particular instances, they can be replicated and tested in other jurisdictions, which is one of the motives behind the I–ADAM initiative.

Pitfalls of Comparative Research

Some words about a few of the pitfalls in cross-national work, alluded to earlier, are in order. A major barrier to be overcome is that of language. Except where research sites share a common language, translation is necessary. Written instruments require not only translation from the original into the second language, but also translation back to the original to verify accuracy. Despite the utmost efforts of the researchers, differences in meaning often creep in that can change the results. Because roughly half of the countries in the I–ADAM project are English speaking, this was not a problem. Nevertheless, it was an issue for the sites in Chile, Malaysia, and the Netherlands. For example, the Chilean report indicates that the interview schedule was translated into Spanish and then given "minor edits and adaptation to the Chilean idiom." It is this kind of adaptation that can introduce imprecision and differences in responses. Language will become an even greater problem if and when countries such as Russia and Ukraine are added as sites, as planned.

Cultural differences in research traditions and methods also have to be reconciled in cross-national studies. The scientific method does not govern everywhere. Sampling can mean different things in different countries. The same is true of access to data, protection of human subjects, and policies with respect to publication of research findings. The Chile report is again a case in point. Referring to the questionnaire, the report says "it was deemed necessary to add as well as remove a few questions on the basis of cultural applicability." Similarly, Australia's report indicates that "changes were made to the instrument based on feedback from local sites, interviewers, and interviewees"—in part because "terminology did not translate." Introducing differences also introduces variability that can threaten validity.

Comparative research is also made difficult by the variation in data reliability and validity across countries. Crimes and criminal justice practices (such as arrests, prosecutions, and convictions) are defined and recorded differently. These difficulties both challenge and discourage researchers and lead to skepticism with respect to the credibility of cross-national research.

These potential problems confronted the I–ADAM project and required the existence, at the outset, of a very compelling argument and rationale for the program, which is stated quite succinctly in the Introduction and Overview chapter. Drug use and the crimes attendant to drug use are global problems. They infect both developed and developing countries. Drugs devastate individual abusers and their families. Drug money feeds transnational organized crime, corrupts governments, and finances civil wars and terrorism. To develop effective policies for confronting these problems, we obviously need to know and learn as much as possible about them. Unfortunately, for many of the reasons outlined earlier, existing drug research makes cross-country comparison difficult or impossible. In many cases, no research existed. The report from Malaysia indicates, for example, that there had never been a study conducted in that country to determine the level of drug and alcohol use among the arrestee population. It was hoped that the ADAM method and protocols could help fill the gaps and overcome incomparability.

Method and Protocols

I–ADAM has systematically attempted to deal with the host of potential problems that are common to comparative research and to ensure uniformity in its approach. The research in the eight countries has proceeded in a parallel way to avoid any history effects. As with the original ADAM, I–ADAM employs two data collection methods: individual interviews and urinalyses. One of the first issues for comparability is the uniformity of the sampling procedure for choosing those subjects from whom to collect data. In accordance with ADAM practice, arrestee participation must be voluntary. Adult male and female arrestees (individuals booked and processed by the police) who are detained long enough to be interviewed, but less than 48 hours, make up the pool from which the sample of volunteers is to be drawn. Sites use probability sampling methods, or if the population is small enough, they interview the entire universe. Alternatively, some sites use convenient sampling methods to reduce the cost of data collection.

A second comparability issue is the need for uniformity with respect to the interviewers and the interview environment. The ADAM protocol stipulates that interviewers should not be law enforcement personnel to avoid any coercion or biasing of responses. All interviewers are carefully trained in the administration of the interview schedule. The interviews themselves are conducted in private or semiprivate settings to encourage confidentiality.

A major issue for uniformity is the interview schedule itself. Dropping and adding questions obviously works against uniformity. The effect of changes is not itself uniform, however. Because all questions in the schedule are not created equal, as long as "core" questions are retained and changes are on the margins, a condition of relative uniformity can be maintained. One effect of alterations in questions is to limit the comparative analyses to just those items that appear in the same manner in all forms.

With respect to urinalysis, many comparability problems are avoided. Assuming the sample has been properly chosen, there is less room for human variation and error in collecting and analyzing urine samples than in conducting the interviews. Five common drugs are tested for all I–ADAM sites: marijuana, cocaine, opiates, amphetamines, and benzodiazepines. Drug testing is a highly reliable method of ascertaining whether a subject has recently used drugs and what drug (or drugs) has been used.

With this background, but also with the reminder that I–ADAM is both limited and in its early stages, we can make some observations about the experiences to date. These observations are, of necessity, impressionistic and suggestive.

Assessment and Observations

The I–ADAM effort described in these pages represents a rare and, in some senses, unique research effort. Because it was adapted from the ADAM foundation, it starts with a tried-and-true methodology. It is also relatively narrowly focused and well defined, thus reducing the risk of incomparability. Unlike other comparative research projects, no considerable time needed to be spent in designing the research (e.g., formulating research questions, sample designs, or data collection and analysis strategies). With the research design already in place, individual country projects could begin as soon as such practical issues as funding and political support were addressed.

The first observation deals with the samples in the different countries. These samples, including those in the U.S., tend to be regarded and even treated as nation specific, when this is clearly not the case. As is pointed out in the U.S. chapter, within-country variability makes any assumptions about population homogeneity untenable. It is generally accepted that due to the variability among and between the 35 sites in the United States, results cannot be generalized to any larger population. This is even a greater problem when results are compared across international borders. Thus, any results will need to be specific and treated with caution.

Considering the sample limitations and variations in the interview schedule, the report from the Netherlands takes what is perhaps the most appropriate perspective on the comparability of data collection at this stage. The Dutch say that "comparisons with I–ADAM data from other countries should perhaps be restricted to hypothesis-forming impressions and should not be used to form conclusions about the drug situation in the Netherlands." This is wise advice. The value of using these reports to develop testable hypotheses and glean other heuristic effects should not be dismissed or downplayed.

There are numerous possibilities to be explored here. For example, South African data show that nearly 80 percent of the offenders arrested for housebreaking tested positively for drugs. Furthermore, the number of housebreakers in South African sites using cocaine was four times that of those committing other crimes. In Australia, the strongest correlation with respect to drug use and offending was between property offenses and opiate use. Scotland found relatively high levels of opiate use among the arrestees tested but much more so among females than males. The U.K. found that many of the measures of drug use correlate strongly with many of the measures of crime. What might be made of all these findings? What are the general patterns of relations between drug use and offending? What is the particular connection here between drugs and housebreaking, and opiates and property offending? Why the gender distinction in Scotland? What are the differences in these respects across countries? These and other questions are stimulated by the current research.

This compendium illustrates that a reasonable beginning has been made. There are problems and there are corrections to be made. Nevertheless, the potential value is clear. The report from England/Wales sums up rather nicely the ultimate value of I–ADAM. Pointing out that these data add to our fundamental knowledge about drug abuse and crime and the way they are linked, the report says, "The data generated by the I–ADAM partnership can be used to compare international trends in drug misuse. The program allows for monitoring changes in drug misuse across countries and over time. This evidence can then be used to inform national and international drug policies."

TABLE OF CONTENTS

Foreword iii
James O. Finckenauer, *National Institute of Justice, Washington, DC, USA*

Executive Summary xi

Introduction and Overview 1
Bruce Taylor, *National Institute of Justice, Washington, DC, USA*

Australia 9
Toni Makkai, *Australian Institute of Criminology, Canberra, Australia*
Doug Johnson, *Australian Institute of Criminology, Canberra, Australia*

Chile 29
Luis Caris, *Chief, Alcohol, Tobacco, and Drug Unit, Ministry of Health, Santiago, Chile*
Bruce Taylor, *National Institute of Justice, Washington, DC, USA*

England and Wales 47
Trevor Bennett, *Institute of Criminology, University of Cambridge, Cambridge, England*

Malaysia 63
Visweswaran Navaratnam, *National Centre for Drug Abuse Research, Universiti Sains Malaysia*
Vicknasingam Balasingam, *National Centre for Drug Abuse Research, Universiti Sains Malaysia*
Hilal Hj. Othman, *National Centre for Drug Abuse Research, Universiti Sains Malaysia*

The Netherlands 75
Peter Klerks, *Dutch Police Academy*
Janine Plaisier, *WODC, Ministry of Justice, The Hague, the Netherlands*
Willemijn Garnier, *Ministry of Justice, The Hague, the Netherlands*

Scotland 97
Neil McKeganey, *University of Glasgow Centre for Drug Misuse Research, Glasgow, Scotland*
Clare Connelly, *University of Glasgow School of Law, Glasgow, Scotland*
John Norrie, *University of Glasgow Robinson Centre for Biostatistics, Glasgow, Scotland*
Janusz Knepil, *Department of Biochemistry, Gartnavel General Hospital, Glasgow, Scotland*

South Africa 117
Elna van Niekerk, *CSIR, Division of Roads and Transport Technology, Pretoria, South Africa*
Charles Parry, *Medical Research Council, Cape Town, South Africa*
Andreas Plüddemann, *Medical Research Council, Cape Town, South Africa*
Antoinette Louw, *Institute for Security Studies, Pretoria, South Africa*
Gina Weir-Smith, *Human Sciences Research Council, Pretoria, South Africa*
D. Locke, *Crime Information Analysis Centre, Pretoria, South Africa*

United States 139
Bruce Taylor, *National Institute of Justice, Washington, DC, USA*

Concluding Thoughts 167
Henry H. Brownstein, *National Institute of Justice, Washington, DC, USA*

Appendix A: Model I–ADAM Survey Instrument 173

Appendix B: Demographic, Economic, and Drug Information by Country 181

Executive Summary

This publication is primarily a progress report on the implementation of the International Arrestee Drug Abuse Monitoring (I–ADAM) program in eight countries and on the challenges they faced. Those countries are Australia, Chile, England, Malaysia, Netherlands, Scotland, South Africa, and the United States. Although some research findings are presented, the principal purpose of this report is to describe the experiences of the countries as they engaged in the process of launching I–ADAM.

Providing the first review of the early stages and status of I–ADAM in participating nations, this volume attempts to identify common problems in developing an I–ADAM site and approaches to surmounting implementation barriers. As noted toward the end of the Introduction and Overview chapter, the sequence of subtopics in each of the eight country chapters is the same, ranging from general information to such issues as obtaining funding, creating a survey instrument, developing research methods, and training.

This report documents that, overall, reasonable progress has been made in implementing the I–ADAM experiment. Problems remain; some adjustments are needed. I–ADAM's impressive potential, however, is undiminished.

International ADAM Program in Brief

I–ADAM is an outgrowth and extension of the Arrestee Drug Abuse Monitoring (ADAM) program, developed and operated by the National Institute of Justice, U.S. Department of Justice. Its origin dating back to the mid-1980s, ADAM tracks trends in the prevalence and types of hardcore drug use—such as cocaine, heroin, marijuana, and methamphetamine—among booked arrestees in 35 sites across the United States. Obtained through a standardized methodology and protocol governing arrestee interviews and urine specimen testing, ADAM provides a series of localized assessments about the U.S. drug abuse problem in the arrestee population that are a central component in studying links between drug use and crime and other social problems.

Aspects of ADAM can be customized to address specific needs at the community level. Thus, data collected through ADAM can also provide a fundamental research and evaluation tool that local analysts, policymakers, and practitioners can use to target drug-related issues of particular interest to the locality, such as evaluations of drug treatment and pretrial diversion programs.

In 1998, the ADAM concept began expanding internationally as the I–ADAM program, envisioned as a research partnership among criminal justice organizations worldwide. Like its namesake, I–ADAM serves the twin functions of research and policy guidance. The program's major goal is to develop a standardized international drug surveillance system—similar survey instruments, sampling, training, and other protocols—that facilitates cross-country comparisons of the prevalence of drug use among arrestees and permits an assessment of the consequences of drug abuse within and across national boundaries.

To the extent that its data identify similar drug problems across national borders, I–ADAM can promote a standard basis for nations to coordinate drug control policies and resources and to improve multilateral cooperation. I–ADAM identification of a growing substance abuse problem in a country's arrestee population can help forecast a potential hot spot for international drug trafficking.

Researchers may also use I–ADAM data in some cases to develop such estimates as the following: the relationship between drugs and crime, sources of illegal income for arrestees, drug dependency, use of substance abuse treatment, drug market dynamics, and certain public health-related consequences of drug abuse. I–ADAM can address topics of local interest (e.g., domestic violence) and international concern (e.g., drug prices).

Comparing Drug Use Across Countries

I–ADAM is a network of researchers from various nations working together to address problems in each of their countries that are unique in some ways yet common in others. One of the themes that emerges from I–ADAM data is that drugs are a global issue. All eight countries have problems. Marijuana in one form or another is the drug that is most commonly used by arrestees throughout the world. In Australia, 50 to 66 percent of all arrestees tested positive for cannabis. In Chile, 30 percent of arrestees tested positive. In South Africa, 36 to 44 percent tested positive.

But some drug problems are unique to particular countries. In South Africa, for example, 19 to 25 percent of arrestees tested positive for mandrax, a blend of methaqualone and antihistamine that is rarely, if ever, found being used by arrestees in other nations.

The most serious problem in each country differs: England and Thailand report that opiates are the main problem. Scotland has a serious problem with injection drugs.

In addition to the similarities and differences in drug use among nations, the nations also report similarities and differences in terms of the problems they have in implementing I–ADAM. For example, in all countries people commit crimes and are arrested and booked daily, night and day. Yet in all countries, it is not possible to conduct interviews 24 hours every day.

The researchers report that in those nations where I–ADAM programs have been implemented or planned, policymakers and practitioners in criminal justice and public health generally express high levels of interest in the program. At the same time, most are experiencing problems identifying and securing adequate funding to support an ongoing program. All must deal with collecting data in a hectic environment, and all must work with local law enforcement and correctional officials to ensure that the program operates in a way that benefits the community without disrupting the ongoing operation of the local jail.

Ultimately, I–ADAM is a fragile multinational network of people who share a common interest in the problems of drugs and are trying to work together to do something about it.

Introduction and Overview

Bruce Taylor, National Institute of Justice

Nations are increasingly recognizing a shared interest in better understanding the causes of crime and drug problems. The problem of drug use and related crime is a global phenomenon. No longer can the drug/crime problem be thought of in isolation or strictly in national terms. The drug trade affects most nations of the world, and the problems that arise because of drugs can be quite severe. In both developing and industrialized countries, problems related to drug abuse include crime, sexually transmitted diseases, accidents, deaths, poverty, unemployment, and decreased productivity. Illicit drugs can be devastating not only at the individual level; they can also corrupt governments and undermine international relations. Despite the growing nature and globalization of the drug problem, the quality and extent of research on this subject has not kept pace. Most drug abuse research to date has been done independently by dozens of countries using different research methods. Unfortunately, the results have not been comparable across countries.

In response to this dearth of international comparative drug research, the National Institute of Justice (NIJ) launched the International Arrestee Drug Abuse Monitoring (I–ADAM) program in 1998, with the first annual I–ADAM conference in Miami. I–ADAM was designed to provide an international research platform to study the relationship between drugs and crime and other social problems. By attempting to identify invariant factors that predict drug use across countries, I–ADAM has the potential to broaden the research community's understanding of the nature of the drug epidemic. More important, I–ADAM is designed to provide a data-driven framework for informing and coordinating global drug control policy. Many recommendations that are made regarding global drug control policy are made without the benefit of empirical research. A fundamental element in designing and planning effective drug control strategies is the presence of extensive and reliable data, which are not available in many countries around the world. I–ADAM is being designed to meet the needs of policymakers by providing a base from which data-driven policy recommendations can be generated. I–ADAM data should make it easier for drug control policies to be coordinated across nations, and improved coordination of drug-control policies could improve the international community's ability to combat the pernicious effects of the global drug trade.

One of the most significant barriers to drug policy research has been funding. Most countries do not have the resources to collect data that are representative of their entire population. However, I–ADAM is an established monitoring system for focusing on the areas of a nation with the worst drug problems (the largest cities), and where people using the most drugs (arrestees) can be found. Also, I–ADAM can provide prevalence estimates for low-base-rate drugs not typically used in the general population (e.g., heroin and cocaine). Finally, the drug use patterns of arrestees are often a good indication of future problems. For example, the U.S. program documented high prevalence rates for cocaine long before the epidemic was detected by drug treatment centers.

Historical Background

In 1987, NIJ created the Drug Use Forecasting (DUF) program to serve as one of the U.S. Government's primary sources of information on drug use among arrestees in cities. DUF became one of the primary research tools on drug use, crime, and related social indicators. In 1997, the DUF program was redesigned and renamed ADAM (Arrestee Drug Abuse Monitoring) to reflect the geographic expansion and increased methodological rigor of the program and its development as both a research and policy platform and a system for locally initiated research on topics identified by sites. At its core, however, the ADAM program preserves DUF's simple concept: interviewing and drug testing arrestees. A component of the new ADAM program was the development of an international drug surveillance system among arrestees. The international element to NIJ's work is relatively new; in the past, NIJ has worked mostly with U.S. Federal, State, and local officials. In 1997, the NIJ International Center was launched as the central organizing body of NIJ's international efforts. Its mission is to stimulate, facilitate, evaluate, and disseminate national and international criminal justice research and information.

With the transition from DUF to ADAM, the development of the I–ADAM program became an important goal. Prior to the formal launching of the I–ADAM program, two countries—Chile and England—started collecting ADAM-like data in 1996.

In April 1998 in Miami, I–ADAM held its first annual conference for representatives from eight nations (Australia, Chile, England, the Netherlands, Panama, Scotland, South Africa, and Uruguay), two international organizations (The Organization of American States and the United Nations International Drug Control Programme), experts in the field of drug surveillance systems, NIJ staff, and other U.S. Federal representatives (the Drug Enforcement Administration and the National Institute on Alcohol Abuse and Alcoholism).

In July 1998, officials from England's Home Office announced the second phase of development for the New England/Wales ADAM program (named NEW–ADAM), which added several new sites and began a time-series design in revisited sites. The second wave of NEW–ADAM data collection began in August 1998; two locations were surveyed in that year (South Norwood in London and Copy Lane in Liverpool).

With the exception of England (1996) and Australia (March 1998), most countries represented at the first I–ADAM meeting in Miami did not have funding for I–ADAM work. In August 1998,

Chile and Scotland secured funds for I–ADAM data collection, and in November 1998, South Africa secured funds for I–ADAM work.

Also during 1998, NIJ began conducting field assessments (in Scotland and England in September 1998, Australia in October 1998, Chile in December 1998, and South Africa in January 1999) and training for interviewer staff (in Australia in January 1999 and Chile in January 1999). In the fall of 1998, the infrastructure for I–ADAM started to take shape, with the development of the core survey and an I–ADAM training program.

At the beginning of 1999, Chile and Australia started data collection; I–ADAM held its second strategic planning meeting in April in Chicago. In addition to those who attended the first meeting, representatives from Malaysia and Taiwan were present at this meeting. In April 1999, NIJ released *Comparing Drug Use Rates of Detained Arrestees in the United States and England*,[1] which represented the first comparative analysis of arrestee drug use rates across two countries and the first publication of the I–ADAM Program.

In mid-1999, three more sites started collecting data (the Netherlands, Scotland, and South Africa), and another country (Spain) contacted NIJ and expressed interest in I–ADAM. NIJ visited Madrid in October 1999 and participated in the Seminarios Internacionales Complutenses. NIJ also participated in an international meeting held by the World Health Organization in Vancouver, Canada, in December 1999 to generate interest in I–ADAM.

In November 1999, NIJ met with other Federal agencies involved in drug policy work, research, treatment, and law enforcement. The dual purposes of the meeting were to inform interested parties about the status, goals and objectives, costs, and future rollout plans for I–ADAM and to start a dialogue among potential Federal or international funding partners and NIJ. One important issue discussed at the meeting was the wisdom of expanding I–ADAM from the eight self-funded countries currently involved to a second generation of I–ADAM sites. At the end of the meeting, a number of representatives expressed interest in I–ADAM on behalf of their organizations and their desire to collaborate with NIJ on developing a program.

In December 1999, the Medical Research Council of South Africa released its first I–ADAM report containing data from three metropolitan sites, and in 2000, a number of countries began publishing the results of their I–ADAM work. Australia released two I–ADAM publications in January and two more in March. Additional publications were released by the Medical Research Council of South Africa in May. Scotland released its first publication in June, and England released its second program publication in July.

Early in 2000, a number of innovative developments occurred in I–ADAM. In February 2000, the first international comparison study of two drug testing kits was completed. This study demonstrated the similarity in results between different drug testing kits; the fact that England is using a different urinalysis kit than the United States does not appear to affect results. In March, data collection for the first-ever national probability sample of arrestees started in South Africa, and the program's first Asian country (Malaysia) began collecting I–ADAM pilot data in May. In early to mid-2000, a number of important I–ADAM-related meetings were held. First, an I–ADAM section was added to the U.S. ADAM annual meeting, and the third annual I–ADAM conference was held in September in Washington, D.C. The third conference was the largest

I–ADAM meeting to date (double the size of the second conference); its more than 50 attendees included 15 representatives from 11 countries (Australia, Chile, England, Malaysia, Russia, Scotland, South Africa, Taiwan, Thailand, Ukraine, and the United States), more than a dozen officials from other U.S. Federal agencies, and researchers and policymakers from a variety of other organizations.

The year 2000 also contained some setbacks. In June, the Netherlands I–ADAM group stopped collecting I–ADAM data.

Early 2001 involved a fair amount of I–ADAM activity. In Australia, three reports were released, and additional funding was approved.[1] In announcing funding of $1.8 million to continue the Australian I–ADAM program, known as DUMA, for a further 2 years, the Honorable Chris Ellison, Minister for Justice and Customs, said: "DUMA testing for the first time provides the criminal justice system with quality data on the drugs-crime link and demonstrates the Government's commitment to evidence-based policy making."

In England, a summary report on the NEW–ADAM program covering July 1999 and April 2000 data collection was issued.[2] The Honorable Bob Ainsworth, Home Office Minister responsible for coordinating the Drugs Strategy, welcomed the publication of the study by saying: "The NEW–ADAM research reveals that the first point in the criminal justice system is picking up a group of particularly prolific drug using offenders. In 1999–2000, this group represented 15 percent of arrestees. Our target is to reduce the size of this group by one quarter in 2005 and by a half in 2008." The summary report presents results of drug use and offending among adult arrestees in eight locations in England and Wales. It sets baselines for the "Communities" component of the national antidrug strategy.

In South Africa, the I–ADAM group published a report on the third phase of data collection for their Three Metros Study, and held a workshop to discuss ADAM findings and implications for policy[3]. A panel of South African experts from justice, corrections, drug research, social services department, offender rehabilitation and police was invited to the workshop to give their views on the drug-crime situation and what was learned from the data.

In the United States, the first release of data collected under the new ADAM sampling scheme and revised survey instrument was made publically available, and ADAM published the "Methodology Guide" and "Data Analytic Guide" by posting the reports on the ADAM Web site. Also, a number of international participants attended the 2001 ADAM Annual Conference held in Miami in June 2001.

Data collection has also proceeded as scheduled in the other I–ADAM participating countries. In 2001, a few other countries (Barbados, Nigeria, and Venezuela) expressed interest in joining the I–ADAM effort. Representatives from Thailand who attended the third I–ADAM annual meeting began preparations to collect I–ADAM data in late 2001 in Chiang Mai city. The three studies are:

1. Makkai, T., and K. McGregor, "Drug Use Monitoring in Australia (DUMA): Annual Report for 2000," Research and Public Policy Series, No. 37, Canberra: Australian Institute of Criminology, 2001; Johnson, D. "Age of Illicit Drug Initiation," Trends and Issues in Crime and Criminal Justice, No. 201, Canberra: Australian Institute of Criminology, 2001; Makkai, T.,

"Drug Use Amongst Police Detainees: Some Comparative Data Testing," Trends and Issues in Crime and Criminal Justice, No. 191, Canberra: Australian Institute of Criminology, 2001.

2. Bennett, T., K. Holloway, and T. Williams, "Drug Use and Offending: Summary Results From the First Year of the NEW–ADAM Research Programme." Home Office Research, Development and Statistics Directorate Research Findings 148. London: Home Office, 2001.

3. Parry, C., A. Louw, and A. Plüddemann, "Drugs and Crime in South Africa: The MRC/ISS 3 Metros Arrestee Study (Phase 3)." Research Brief. Pretoria: Medical Research Council and Institute for Security Studies.

Description of Program

The I–ADAM Program is an international partnership of government-sponsored research organizations. All participating countries/jurisdictions are being operated through local/national funds. Countries participate in I–ADAM to stay current on the latest scientific methods of arrestee drug surveillance survey work, take part in comparative data analytic projects on cross-national differences in drug use, and receive technical assistance on the basic operation of a data collection system.

I–ADAM's development is important because the existing drug surveillance systems around the globe, in many cases, are not compatible. Therefore, post-hoc comparisons across countries with independently designed systems are very difficult. The existing general population household surveys found in some countries use very different measures of drug use and were not designed for multinational comparisons. From its inception, I–ADAM has been designed to be a standardized international surveillance system (with similar instruments, sampling, training, and other protocols).

I–ADAM serves two functions: research and policy guidance. Research might show, for example, that certain market conditions must exist for particular drug epidemics to thrive. The findings could then be used by countries that are not experiencing the epidemic to plan prevention efforts. It is hoped that, in the future, I–ADAM can help form a standard basis for nations to coordinate drug control policies. A growing substance abuse problem in a country's arrestee population can help forecast a potential hot spot for international drug trafficking. Through development of addendums, I–ADAM can be used to help estimate the characteristics of drug markets around the world; such data should prove useful for coordinating international drug control efforts.

I–ADAM is a standardized international drug surveillance system that for the first time provides researchers with a platform to compare the prevalence of drug use among arrestees in different nations and allow them to assess the consequences of drug abuse within and across national boundaries. I–ADAM has three main components: voluntary, anonymous, and confidential interviewing; similar data collection methods; and standardized measures.

Voluntary, anonymous, and confidential interviewing

At I–ADAM data collection sites, trained interviewers (who are not in law enforcement) conduct individual interviews with detained arrestees and collect voluntary and anonymous urine specimens from interviewees. Every I–ADAM site provides a private or semiprivate interview environment that is conducive to open, valid, and reliable responses by participants.

Similar data collection methods

Each I–ADAM site uses similar eligibility criteria for selecting study participants. Site staff collect data from male and female booked/processed arrestees detained long enough to be interviewed (but less than 48 hours). Study participants are generally interviewed before they have seen a magistrate or judge, but they have access to legal counsel if they wish to ask about the voluntary nature of the study.

Data collection from juvenile males and females is optional. Each I–ADAM site attempts to obtain a sample size large enough to provide a reasonable level of statistical precision where all the main age and gender groups are represented in sufficient numbers. Typically, the site schedules about 2 to 3 weeks of quarterly interviewing at the jail/lockup. Most I–ADAM sites collect data from 150 adult male and 75 adult female arrestees each quarter to reach a total of 900 arrestees annually. I–ADAM sites aim to use a definable study/catchment area and to understand the representativeness of their data. Ultimately, probability-based sampling methods will be adopted at each I–ADAM site. At I–ADAM sites in England and Scotland, the entire universe of eligible arrestees is interviewed at selected study sites during a selected time period. When the universe of cases is interviewed, the problem of complicated sampling schemes is avoided, but the cost of data collection increases dramatically. The sites that have adopted population-based data collection schemes have used this method because of the small number of arrests made in their jurisdictions. By interviewing all of the eligible cases, these sites are able to maximize the number of interviews that can be conducted in the shortest period of time, thereby reducing the burden of data collection to a few weeks per quarter.

Standardized measures

In collaboration with NIJ, international sites determine the type and number of drugs in the drug test panel. Currently, at least five common drugs (marijuana, cocaine, heroin/opiates, amphetamines, and benzodiazepines) are being tested by all the I–ADAM sites. Also, NIJ has developed a core I–ADAM survey instrument in consultation with the other I–ADAM sites, and the agreed-on core survey is being implemented by all the active sites. The development of a revised core I–ADAM survey, which might include such new topical areas as drug dependency, criminal history, and drug markets, is under consideration. At a later point, common addendum surveys will be developed for special topical areas (e.g., domestic violence).

Organization of This Compendium

The chapters in this report provide a review of the developmental stages of I–ADAM in participating countries, barriers to development, national stakeholders involved in the development of I–ADAM, research methods used in I–ADAM, and analytic results (if data are available). This report does not involve direct statistical testing of differences across countries; instead, it is intended to be descriptive. The report attempts to identify common problems across countries in developing an I–ADAM site and approaches to resolving those problems. Each country provides a summary of its I–ADAM results and likely policy implications or applications of its data. Also, site description tables covering sociodemographic characteristics, economic information, and drug use data are included in the appendix for each I–ADAM country.

All country chapters follow the same basic outline. The first three sections of each chapter provide some broad information about the participating countries. Here, the attempt was made to provide some context for I–ADAM work in terms of the general sociodemographic conditions present in each country, the basic operation of the criminal justice system within which I–ADAM needs to fit, and the status of drug control policy. Section I covers national census information on each country. A basic demographic profile is provided, including statistics on population size; geographic size; ethnic, gender, and age distribution; unemployment, poverty, and literacy rates; income; gross national product; percent living in urban conditions; birth rate; life expectancy; population growth; mortality rates; and gender composition. Section II provides a description of the country's criminal justice system, the arrest process, and the detention system. Prior research with DUF (and now ADAM) has shown that the study of arrestees is important because they typically represent the hardcore population of drug users and are the group consuming the majority of illicit drugs.

Section III examines the general policies in place to deal with drugs in each participating country. A fair amount of variability exists within I–ADAM on the issue of drug control policy. For example, the Netherlands has a policy of harm reduction, or a medical model. Therefore, consumption of certain small amounts of drugs is tolerated by law enforcement in the Netherlands. Malaysia, however, has a strict zero-tolerance policy for drugs, and the death penalty exists for certain drug offenses. The overview of a country's drug control policy helps to explain the drug use of its population.

Section IV covers the developmental history of I–ADAM in each country. For most participating countries, this history is fairly recent—beginning around 1998—and the programs have developed quite rapidly. Section V examines the goals and objectives of I–ADAM. It is important to understand the local/national focus of I–ADAM and that, without the achievement of local goals, little room for international work exists. Most participating countries have not formally stated their goals and objectives, but the authors identify at least some of the broad aims of the program. Section VI documents the difficult work of getting an I–ADAM site operational and is meant to help interested countries start a program. Issues reviewed in this section include:

- Obtaining funding.
- Deciding to conduct in-house research or to contract it out.
- Negotiating access to the lockup facilities where arrestees are held.

- Learning about the catchment area for the project and how arrestees are processed therein.
- Handling urine specimens.
- Creating a survey instrument that is culturally applicable for the local arrestee population.
- Designing a survey that conforms to I–ADAM core standards but retains its utility for local purposes.

Section VII reviews the numerous barriers that exist to conducting this type of work, such as obtaining adequate funding and obtaining a large enough sample to allow sufficient statistical precision. Given the importance of securing political support for I–ADAM to sustain the program, section VIII discusses the national stakeholders involved. Section IX reviews the site-specific research methods used in the program. This section pays special attention to detailing the sampling scheme and related data collection protocols. Section X reviews training issues. NIJ has developed a training curriculum and related materials and made it available to all participating I–ADAM jurisdictions. NIJ and its contractors have also visited each of the countries to assist and/or participate in the development of local training programs. This section identifies difficulties that have been encountered in training I–ADAM interviewers and areas that need to be focused on in the future. The final sections review the main data analytic results of I–ADAM in each country (Malaysia did not have findings at the time this compendium was written), the policy implications of I–ADAM, and various examples of I–ADAM data applications.

Note

1. Taylor, B., and T. Bennett, *Comparing Drug Use Rates of Detained Arrestees in the United States and England,* Washington, DC: U.S. Department of Justice, National Institute of Justice, 1999, NCJ 175052.

AUSTRALIA

Toni Makkai
Doug Johnson

Toni Makkai is director of research at the Australian Institute of Criminology and program director of the Drug Use Monitoring in Australia (DUMA) project. She has held university teaching and research appointments in England and Australia. **Doug Johnson** is currently a project officer with the Commonwealth Attorney General. Prior to this, he was a member of the DUMA project team at the Australian Institute of Criminology.

DUMA is funded under a 3-year grant from the Commonwealth's National Illicit Drug Strategy. The data used here were collected for the Australian Institute of Criminology's Drug Use Monitoring in Australia project by the National Drug Research Institute at the Curtin University of Technology, Hauritz & Associates Pty Ltd, and Forsythe Consultants Pty Ltd, with the assistance of the Queensland, New South Wales, and Western Australian Police Services. Neither the collectors nor the police services bear any responsibility for the analyses or interpretations presented here.

The opinions expressed here are those of the authors and do not necessarily reflect those of the Australian Institute of Criminology.

I. Overview of Demographic Profile of Nation

Australia has one of the most stable and conservative democracies in the world. The official Head of State, the Queen of England, and the Westminster style of government are evidence of its British heritage. Parliament consists of two chambers, the House of Representatives and the Senate. Voting is along party lines, and the party that captures a majority of seats in the House of Representatives—or that can assemble a coalition with a majority of seats—is given the opportunity to form a government. The majority party's leader becomes the Prime Minister. The Australian political system is a federal system in which the States have significant power relative to the Commonwealth Government. The country comprises six States and two Territories. Each State has its own Parliament and Premier, and each guards its sovereignty jealously. The Territories also have their own Parliaments and Chief Ministers, but they have less sovereignty over their own internal affairs.

Australia is at once one of the world's most sparsely populated countries and one of its most urbanized. A population of 19.1 million, 1/14th that of the United States, occupies a landmass of 7,692,030 square kilometers, nearly the size of the United States. At the same time, 80 percent of Australia's population is clustered in a few large coastal cities at the edge of an otherwise virtually uninhabited continent. Characteristics of the population, including population growth, life expectancy, birth and death rates, and literacy (see exhibit 1) are in keeping with those of other industrialized, Western countries.[1] Australia's population enjoys a high standard of living, as shown in exhibit 2.

Besides its geographic distribution, the other distinguishing feature of Australia's population is the role that immigration has played in its growth. Although Australian-born individuals contributed two-thirds of the increase in population between 1901 and 1998, by June 1999, 24 percent of the population was born overseas. Moreover, although the population is still heavily Anglo-Celtic in background, the sources of immigration have diversified since World War II.

Drug and Crime Statistics

The primary sources of drug statistics in Australia are the National Drug Strategy Household Survey on Drug Use, school-based surveys, and the annual Illicit Drug Reporting System (IDRS), which includes surveys of injecting drug users (IDUs). The household and school surveys are random samples conducted every 2–3 years and are based on self-report data.[2] The most recent surveys show upward trends in the use of illicit drugs. As shown in the drug arrest statistics (see exhibit 3), cannabis remains the most popular illicit drug, followed by heroin and amphetamines. IDRS is conducted annually; IDU surveys are convenience samples that use self-report data. The IDU surveys show that heroin and amphetamines are the most popular illicit drugs among IDUs.

The Australian Bureau of Criminal Intelligence collects annual data on drug-related issues (exhibit 3), price, and purity. The average drug purity in 1998–99 was 65 percent for heroin, 51 percent for cocaine, and 17 percent for methamphetamines.[3] Prices vary across jurisdictions. In January–March 1999, the price of a cap (0.1–0.3 g) of heroin was estimated to range from $A25 in parts of New South Wales to $A100[4] in the Northern Territory.

Exhibit 1. General Demographic Information for Australia

Geographic area	7,692,030 km²
Estimated population	19.1 million
Population by place of origin (1996–99)*	24% overseas born, 73.9% Australian born/nonindigenous, 2.1% indigenous
Gender distribution[†]	50.2% female; 49.8% male
Age distribution[‡]	0–9 years, 13.7 %; 10–14 years, 6.9%; 15–19 years, 7.0%; 20–29 years, 15.0 %; 30–39 years, 15.3%; 40–54 years, 20.9%; 55–64 years, 8.8%; ≥65 years, 12.3%
Life expectancy (at birth for 1996–98 cohort)[§]	Male, 75.2 years; female, 81.9 years
Population growth rate**	1.1%
Total fertility rate	1.8%
Death rate	5.89 per 1,000
Literacy rate[††]	Level 1, 17%; level 2, 28%; level 3, 38%;0 levels 4 and 5, 17%
Political subdivisions	6 States: Queensland, Victoria, New South Wales, Western Australia, South Australia, Tasmania; 2 Territories: Northern Territory, Australian Capital Territory

*As of June 1999. In 1947, 81 percent of overseas-born Australians were from the main English-speaking countries (United Kingdom, Ireland, New Zealand, South Africa, Canada, and the United States); by 1998, the percentage of overseas-born Australians from these countries had dropped to 39 percent. In addition, the 1996 census shows that 27 percent of persons born in Australia had at least one parent born overseas. Australian Bureau of Statistics, *Migration, Australia 1998–99,* Canberra: Australian Bureau of Statistics, 2000. ABS 3412.0; and Australian Bureau of Statistics, *Year Book Australia 2001,* Canberra: Australian Bureau of Statistics, 2001, ABS 1301.0, tables 5.34 and 5.35.

[†]Australian Bureau of Statistics, *Australian Demographic Statistics,* Canberra: Australian Bureau of Statistics, March quarter 2000, ABS 3101.0.

[‡]Australian Bureau of Statistics, *Population by Age and Sex, Australian States and Territories,* Canberra: Australian Bureau of Statistics, 1999, ABS 3201.0.

[§]Australian Bureau of Statistics, *Year Book Australia 2001,* Canberra: Australian Bureau of Statistics, 2001, ABS 1301.0, graph 5.25 and table 5.26; and Australian Bureau of Statistics, *Deaths: Australia 1998,* Canberra: Australian Bureau of Statistics, 1998, ABS 3302.0.

**For the year ending March 2000. Australian Bureau of Statistics, *Australian Demographic Statistics,* Canberra: Australian Bureau of Statistics, March quarter 2000: ABS 3101.0.

[††]These figures come from a literacy survey conducted by the Australian Bureau of Statistics in 1996 for the second phase of the International Adult Literacy Survey jointly sponsored by the Organisation for Economic Cooperation and Development and Statistics Canada. See Organisation for Economic Cooperation and Development, *Education at a Glance,* Paris: Organisation for Economic Cooperation and Development, 2000, table A2.3a. Generally, levels 1 and 2 represent very poor and poor literacy; level 3 represents the minimum level for functioning in an advanced society (high school graduate or college-entry level); and levels 4 and 5 represent command of higher order information processing skills. For a more detailed discussion of the five levels of literacy, see Organisation for Economic Cooperation and Development and Statistics Canada, *Literacy in the Information Age: Final Report of the International Adult Literacy Survey,* Paris: Organisation for Economic Cooperation and Development, and Ottawa: Minister of Industry, Canada, 2000. For more specific information about the Australian survey, see Australian Bureau of Statistics, *Aspects of Literacy: Profiles and Perceptions, Australia, 1996,* Canberra: Australian Bureau of Statistics, 1997; and Australian Bureau of Statistics, *Aspects of Literacy: Assessed Skill Levels, Australia, 1996,* Canberra: Australian Bureau of Statistics, 1997.

The last national crime victimization survey was undertaken in 1998. Data from this collection found that the prevalence rate for all offenses against households was 9.0 percent and for all offenses against persons was 4.8 percent. The most common crime was breaking and entering, with a prevalence rate of 7.6 percent. The survey found that those who were assaulted during a 12-month period were more likely to be repeat victims than were victims of other crimes. Forty-six percent of assault victims had experienced two or more assaults during the previous 12 months.

Since 1993, national crime statistics have been collected on seven major crimes recorded by police in Australia: homicide, assault, sexual assault, robbery, unlawful entry with intent, motor vehicle

Exhibit 2. Economic Information for Australia

Gross domestic product (GDP) (March quarter 2001)	$159.0 billion
GDP growth rate (December 2000–March 2001)	0.2% trend, 1.1% seasonally adjusted
GDP growth rate (March 2000–March 2001)	1.6% trend, 2.1% seasonally adjusted
Average weekly earnings (February 2001)	$A655.30, all employees, $A844.70, full-time adult total earnings
Growth in average weekly earnings (November 2000–February 2001)	0.8% all employees, 0.6% full-time adult total earnings
Growth in average weekly earnings (February 2000–February 2001)	5.0% all employees, 4.0% full-time adult total earnings
After-housing poverty rate (1989–90)*	13.0%
Unemployment rate (April 2001)	6.8%

*See Saunders, Peter, "Poverty and Deprivation in Australia," in *Year Book Australia 1996*, Canberra: Australian Bureau of Statistics, 1996. According to statistics cited by the Australian Council on Social Service, the after-housing poverty rate in Australia increased from 8.2 percent in 1973 to 11.5 percent in 1996. See King, A., "Income Poverty Since the Early 1970s," in *Australian Poverty Then and Now*, ed. R. Fincher and J. Nieuwenhuysen, Melbourne: Melbourne University Press, 1998, cited in Australian Council of Social Service, "ACOSS Submission to the Reference Group on Review of the Welfare System," paper 103, Strawberry Hills, Australia: Australian Council of Social Service, 1999: 4.

Sources: Australian Bureau of Statistics, *Australian National Accounts: National Income, Expenditure, and Production,* Canberra: Australian Bureau of Statistics, 2001: ABS 5206.0; Australian Bureau of Statistics, *Average Weekly Earnings: States and Australia,* Canberra: Australian Bureau of Statistics, 2001: ABS 6302.0; and Australian Bureau of Statistics, *Labour Force—Unemployment Rate—Australia—Monthly,* Canberra: Australian Bureau of Statistics, 2001, ABS 6291.0.40.001, table 91.

Exhibit 3. Drug Arrests in Australia, 1998–99

Total drug arrests	83,5241*
Male	68,193
Female	14,212
Consumers	65,836
Providers	17,688
Percentage of arrests by type of drug	
Cannabis	69.6%
Heroin	17.2%
Amphetamines	7.9%
Cocaine	0.7%
Hallucinogens	0.7%
Steroids	0.1%
Other	3.8%

*Male and female arrests do not equal total drug arrests because some data are missing.

Source: Australian Bureau of Criminal Intelligence, *Australian Illicit Drug Report 1998–99,* Canberra: Australia Bureau of Criminal Intelligence, 2000, tables 11.1 and 11.16.

theft, and other theft. The most common crime recorded by police is "other theft," which includes pickpocketing, bag snatching, and stock stealing. This category accounts for slightly less than half of the total recorded crimes. The rates of assaults increased from 1993 to 2000. The rates of property offenses and robberies also increased from 1993 to 1998, decreased in 1999, and appear to have increased again in 2000. In 1999, 342 murders were recorded, which represented a rate of 18 per 1 million population; murders decreased by 12 percent in 2000 to 302.[5]

II. Description of Criminal Justice System, Arrest Process, and Detention System

Australia's federal system of government has implications for its criminal justice system. The States and Territories administer criminal justice, but "the components and processes of criminal justice are essentially the same across Australia."[6] Australia is served by eight police forces: one in each State and the Northern Territory, and the Australian Federal Police, which also polices the Australian Capital Territory. The Drug Use Monitoring in Australia (DUMA) project operates in three States: Queensland, Western Australia, and New South Wales.

Exhibit 4 illustrates the usual arrest process for offenders. Once an offense comes to the attention of the police, the officer can decide whether to investigate further. If the officer proceeds with the investigation, three options are open: The suspect can be issued a "notice to attend" or its equivalent, given a summons, or arrested. A notice to attend is usually given for relatively minor offenses. A summons is given for civil offenses such as traffic incidents or tax matters. In both these

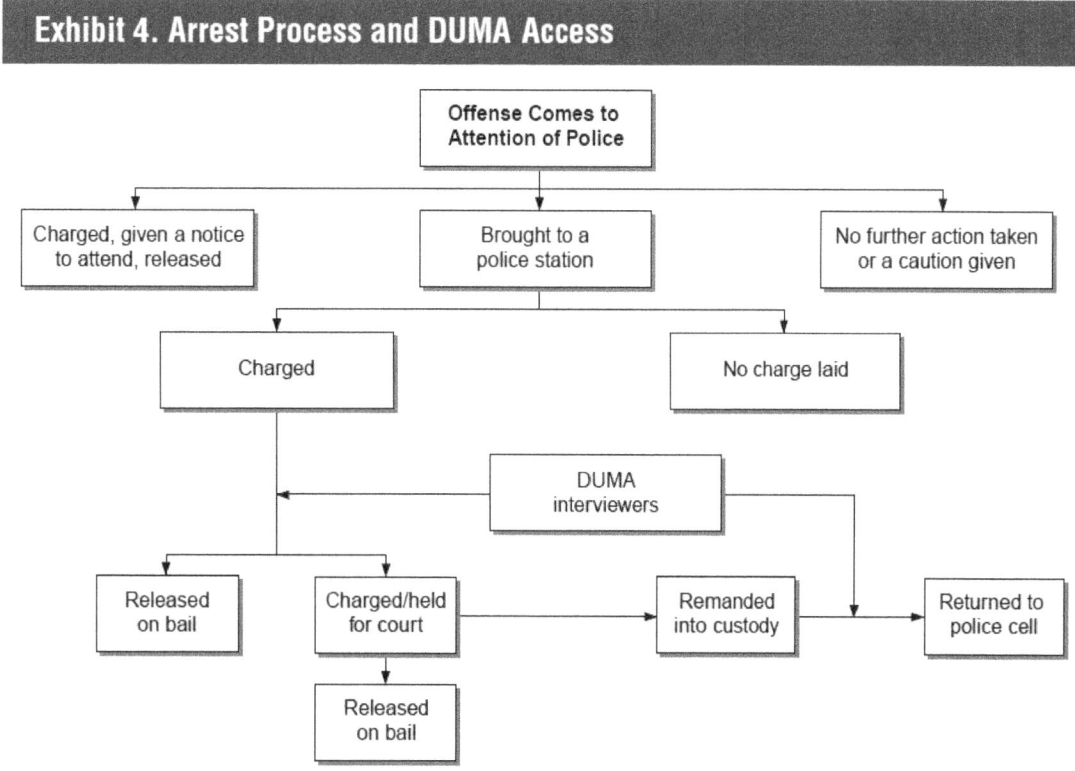

Exhibit 4. Arrest Process and DUMA Access

instances, the offender is usually not brought to the police station. If the matter is more serious, the officer can either arrest the suspect immediately and bring him or her to the police station or ask the suspect to accompany him or her to the station for further investigation. Once at the police station, the arresting officer can decide to continue with a formal charging process or release the suspect from custody. If the suspect is charged, the seriousness of the charge will determine whether the suspect is to be held in custody for a magistrate's hearing. If the charge is not serious, the suspect is released on bail following completion of formal charges. In the magistrate's court, the suspect can either be given bail and released or refused bail and remanded into custody.

DUMA staff cannot interview a detainee until the arrest process is complete. In some sites, if the offender is refused bail, he or she is then transferred to corrective services. DUMA personnel must interview the detainee before the transfer. Some sites also have a maximum detention time. In New South Wales, the police can detain a suspect for up to 4 hours for investigative purposes (an application can be made for another 8 hours). After that time, the suspect must be charged or released. DUMA staff must gain access to the detainee before this deadline. There is a last window of opportunity to interview those people charged and then remanded into custody by the court; if remand is within 48 hours after arrest, then data collection can proceed.

Australia has State and Federal courts. Seventy-nine percent of court hearings in 1996–97 were for criminal matters, which State and Territory magistrates' courts usually handle. These courts have no jury and normally cannot impose a sentence longer than 2 years. Most defendants in the magistrates' courts are found guilty; the majority of defendants are male. The district or higher court hears more serious criminal matters.

Defendants who cannot afford their own solicitors can apply for legal assistance, but they have no constitutional right to representation. On conviction, the courts can impose a range of measures, including a fine, suspended sentence, community-based supervision order, home detention, periodic detention, and incarceration. In 1997–98, 74,810 persons were sentenced in Australia. Of these, 25 percent were placed in a community corrections program, 25 percent were given a prison sentence, and 2 percent were given periodic detention.[7] The rate of imprisonment in Australia of persons older than age 16 has increased significantly in the past two decades, rising 52 percent from 91.6 per 100,000 population in 1983 to 139.2 per 100,000 population in 1998.

III. Status of National Drug Policy

The drug of choice in Australian society is alcohol, and, although the country's temperance movement was active in the late 19th and early 20th centuries, alcohol was never prohibited.[8] Before Federation in 1901, opium and cocaine were legal and as widely available as alcohol and tobacco. At the turn of the 20th century, considerable concern grew about the use of opium, particularly by ethnic Chinese. A combination of factors caused the States to introduce legislation to ban opium use. During the early 1900s, laws were passed to ban over-the-counter medications, including heroin. Those drugs no longer could be sold without a doctor's prescription. A policy of drug maintenance under the control of general practitioners continued until the early 1950s. Concurrently, an international outcry about high levels of legal heroin use led to a Federal ban on the importation of heroin, and Australian States banned the manufacture of heroin within their jurisdictions.

By the early 1980s, the first cases of AIDS were diagnosed, civil liberties were increasingly being eroded, the profits from organized crime were growing exponentially, and public officials were being corrupted by large sums of money from the drug trade. In 1983, the Federal Government introduced a national harm minimization strategy that was most clearly defined in the 1993–97 National Drug Strategic Plan:

> Harm minimization is an approach that aims to reduce the adverse health, social, and economic consequences of alcohol and other drugs by minimizing or limiting the harms and hazards of drug use for both the community and the individual without necessarily eliminating use.[9]

Successive governments have continued to endorse this policy, but at first, law enforcement was not supportive. By the 1990s, law enforcement agencies were committed to harm minimization, at least at the policy level. Throughout the 1990s, a range of drug and crime indicators suggested that illicit drug use and related criminal activity were increasing. The Federal Government responded in 1996 with the launch of the National Illicit Drug Strategy (NIDS), which allocated more money to both supply and demand reduction strategies.

IV. Background History to DUMA

Throughout the 1990s in Australia, deaths due to opioid overdose increased at an alarming rate; illicit drugs, particularly heroin, were widely available; and open drug markets started to appear in some cities. Crime, particularly property crime, was increasing. In 1997, the Federal Government reinvigorated its commitment to a national drug policy by allocating more money for law enforcement, treatment, and education. Given the severity of the problem and the link between illicit drugs and crime, the Australian Institute of Criminology (AIC) successfully applied for funds to conduct the Drug Use Monitoring in Australia (DUMA) pilot project. The project began collecting data in January 1999 and continued under current funding arrangements until December 2001.

V. Goals and Objectives of DUMA

As a pilot study, DUMA's primary purpose is to apply the U.S. Arrestee Drug Abuse Monitoring (ADAM) methodology and questionnaire to Australia; its secondary purpose is to begin collecting credible data on drugs and crime. The project specifically aims to determine whether the ADAM methodology can provide the following:

- Reliable drug prevalence data on detainees.
- Aggregated data to local and national law enforcement and treatment agencies for monitoring, evaluation, and assessment purposes.
- Timely and high-quality data to inform policy on drugs and crime.[10]

VI. Getting Started

A. Obtaining Funding

NIDS provided new funds for drug policy, half of which was directed toward supply interdiction efforts. AIC obtained $1.6 million in funding for DUMA over 3 years from NIDS's supply-side funds, which represents 0.76 percent of the total money allocated to supply-side intervention strategies.

B. Contracting for Data Collection

AIC does not collect DUMA data itself. Instead, a partnership with local police and local universities or research companies operates at each site. Memorandums of understanding govern the working relationship between AIC and the data collectors. Each site has its own advisory or steering committee that deals with local strategic issues. Participating State police services are well represented on the local committees.

C. Facility Access

Facility access must be obtained at two levels. At the higher level, the State police commissioner in the relevant jurisdiction must give permission for access. At the second level, local watchhouse personnel must cooperate with the data collection team. Interviews take place at each site usually over a 3-week period every 3 months. Fieldwork began in January 1999 at the Queensland (Southport watchhouse) and Western Australia (East Perth lockup) sites and in the third quarter of 1999 at the two New South Wales sites in Sydney (Bankstown and Parramatta). Exhibit 5 shows the periods when fieldwork was undertaken. Start dates at each site varied according to local conditions, but they were generally within a few weeks of each other. The collection of data at the two New South Wales sites—one after the other rather than at the same time—complicated matters.

Only adult males were interviewed in the first quarter of 1999 in the Queensland and Western Australia sites. Collection was expanded to include adult females in Queensland in the second quarter of 1999 and in Western Australia in the third quarter of 1999. In New South Wales, adult males, adult females, and juveniles were interviewed from the start. The legal definition of an adult is age 17 in Queensland and age 18 in Western Australia and New South Wales.

DUMA interviewers cannot be law enforcement officers or court personnel. Given Australia's size, interviewers must be recruited locally. Thirty-one interviewers were used across all sites in 1999. The average age of interviewers was 33. Twenty of the 31 interviewers were female, and 23 were Australian born. Interviewers were often postgraduate students enrolled in psychology, drug research, or criminal justice programs. All interviewers were required to undergo training before each data collection. As part of the training, they engaged in role playing and discussed their error reports from the previous collection.

Exhibit 5. Fieldwork Information for DUMA Pilot Data Collection

Quarter	Site	Period	Hours in facility	Number approached	Number interviewed	Specimens collected
Q1	Western Australia	01/18/99–01/31/99	248.0	138	116	76
	Queensland	01/15/99–02/07/99	160.5	99	93	80
Q2	Western Australia	04/07/99–04/25/99	258.0	162	134	82
	Queensland	04/15/99–05/12/99	168.5	170	158	133
Q3	New South Wales* (Bankstown)	06/28/99–07/16/99	246.0	108	91	45
	Western Australia	07/02/99–07/25/99	416.0	158	124	78
	New South Wales* (Parramatta)	06/07/99–06/27/99	312.0	110	94	43
	Queensland	07/15/99–08/11/99	193.0	131	131	106
Q4	New South Wales* (Bankstown)	10/11/99–11/01/99	298.5	104	87	52
	Western Australia	11/08/99–11/28/99	344.0	139	105	69
	New South Wales* (Parramatta)	11/01/99–11/21/99	257.5	137	107	74
	Queensland	10/08/99–11/04/99	183.3	185	162	140
Total	All sites	1999	3,085.3	1,641	1,402	978

* Numbers for the New South Wales sites include juveniles.

After police processing, DUMA interviewers approached detainees and asked them to participate in a confidential and voluntary research project. Detainees were shown a statement describing the study; interviewers read the statement to those with reading difficulties. Interviewers pointed out that the detainees did not have to take part in the interview if they did not want to, they did not have to answer any questions that they did not want to, and they could stop the interview and leave at any time. Detainees were then asked whether they agreed to take part in the study.

In 1999, seven people started the interview and chose not to complete it. Detainees could also choose to complete the interview but not provide a urine specimen. Three hundred and fifty detainees (25 percent) who completed interviews in 1999 chose not to provide urine specimens; another 66 (5 percent) tried to produce a specimen but could not. At some sites, detainees were offered food or coffee or tea to thank them for taking part.

D. Knowing the Catchment Area

Although the sites are referred to by the name of the area where the site is located, the catchment area may not necessarily reflect the city boundaries. A number of differences were observed among the four sites. The size of the population in the catchment areas varies greatly (see exhibit 6). This reflects the nature of each site. The East Perth site is a large, old facility designed to take detainees from a large portion of the city. Many police stations in the city have the capacity to handle only a few detainees; the East Perth facility accommodates detainees who are to be held

Exhibit 6. Profile of Geographical Areas Surrounding DUMA Sites

Criterion	Queensland	Western Australia	New South Wales (Bankstown)	New South Wales (Parramatta)
Number of people in catchment area	332,952	294,957	158,358	58,962
Mean age of population (years)	39	38	37	36
Male (%)	49	48	49	50
Australian born (%)	69	62	63	61
Employed (%)	86	92	95	91
Left school at age 14 or younger (%)	16	12	20	11
Never married (%)	28	32	29	28

Source: Australian Bureau of Statistics, 1996 Census, Canberra: Australian Bureau of Statistics, 1997.

overnight if a local police station is not open 24 hours a day or does not have a holding cell available. Sydney has no large central facility for detainees. Local police stations handle their own detainees and take them from other parts of the city only in emergencies. It is rare for detainees to be held for longer than 4 hours in these facilities; if they are to be held overnight, they are transferred to Corrective Services. The Southport watchhouse caters to the entire southeastern region of Queensland. It is a modern facility with the capacity to hold detainees overnight and serves the local magistrate's court.

Police districts do not always overlap with census districts, so it is difficult to determine the precise characteristics of the catchment areas using census data. Postal codes and local census districts were used to match the police districts as closely as possible. Exhibit 6 provides basic census data on each catchment area. East Perth, Bankstown, and Parramatta differ little in terms of the mean age of the population, the percentage of males, and the percentage of Australian-born people. The catchment population in Southport has a slightly smaller proportion of people with jobs, and both Bankstown and Southport have a higher percentage of people who left school at age 14 or younger. The percentage who have never married ranges from 28 to 32 at all sites.

E. Drug Testing Issues

All drug testing for the DUMA project was conducted at Pacific Laboratory Medical Services, Northern Sydney Area Health Service, in Sydney. The laboratory is accredited to the Australian Standards AS3408–1995, and it consistently seeks to improve the sensitivity of its tests. In 2000, the EMIT™ (enzyme multiplied immunoassay testing) assay for methadone and amphetamines will be changed to CEDIA®. CEDIA shows fewer false positives due to proprietary medications, and its reagent is more sensitive to MDMA (Ecstasy or XTC) and methylamphetamine.

Urine is routinely tested for six classes of drugs: cannabis, opiates, methadone, cocaine, amphetamines, and benzodiazepines. A test is deemed positive when the drug or its metabolites are detected at the cutoff levels prescribed in Australian Standard AS3408–1995.[11] Confirmatory tests are undertaken when a sample tests positive for opiates, amphetamines, or benzodiazepines.

F. Instrument Issues

Cultural applicability

DUMA has attempted to keep the core questionnaire as close as possible to the U.S. ADAM questionnaire. Throughout 1999, however, changes were made to the instrument based on feedback from local sites, interviewers, and interviewees. In some cases, terminology did not translate or distinctions (i.e., summary/indictable) were not appropriate for some sites. As a result of interviewer complaints about the length of the drug grid, it was broken down into separate grids. In addition, interviewers reported that they were skeptical about detainees' responses to questions related to money earned from illegal activities. As a result, these questions have been moved toward the end of the questionnaire in the hope that initial doubts about the confidentiality of the process will recede as rapport develops between the interviewer and the interviewee.

Standardization versus local utility

Each site has local concerns, and the pressure to include other questions is always present. The means of overcoming this pressure is to encourage sites to develop local addenda that can be used in their jurisdictions to answer specific policy questions.

VII. Barriers to Developing DUMA

Although AIC secured money for the DUMA pilot project, ongoing funding is by no means certain. The five potential barriers to the long-term development of DUMA are—

- Securing ongoing funding and cultivating greater support at the policy level.
- Further promoting cooperation and dedication of resources by police at the local site.
- Maintaining a well-trained interview team when the work is intermittent.
- Recognizing and solving ethical problems when dealing with a vulnerable population.
- Addressing legal issues in which neither the interviewer nor the detainee has protections such as those provided by the statutory and regulatory confidentiality requirements stipulated under U.S. legislation.[12]

VIII. National Stakeholders Involved in the Development of DUMA

DUMA is a partnership among AIC, local police, and researchers. Exhibit 7 indicates the general arrangements by which the DUMA project is overseen. The project has been funded by NIDS to run pilot sites in three jurisdictions (Queensland, Western Australia, and New South Wales) over 3 years. It is affiliated with the International Arrestee Drug Abuse Monitoring (I–ADAM) program and has a scientific advisory committee with whom AIC consults concerning technical and methodological issues.

Exhibit 7. Sponsoring Organizations, Participants, and Stakeholders in DUMA

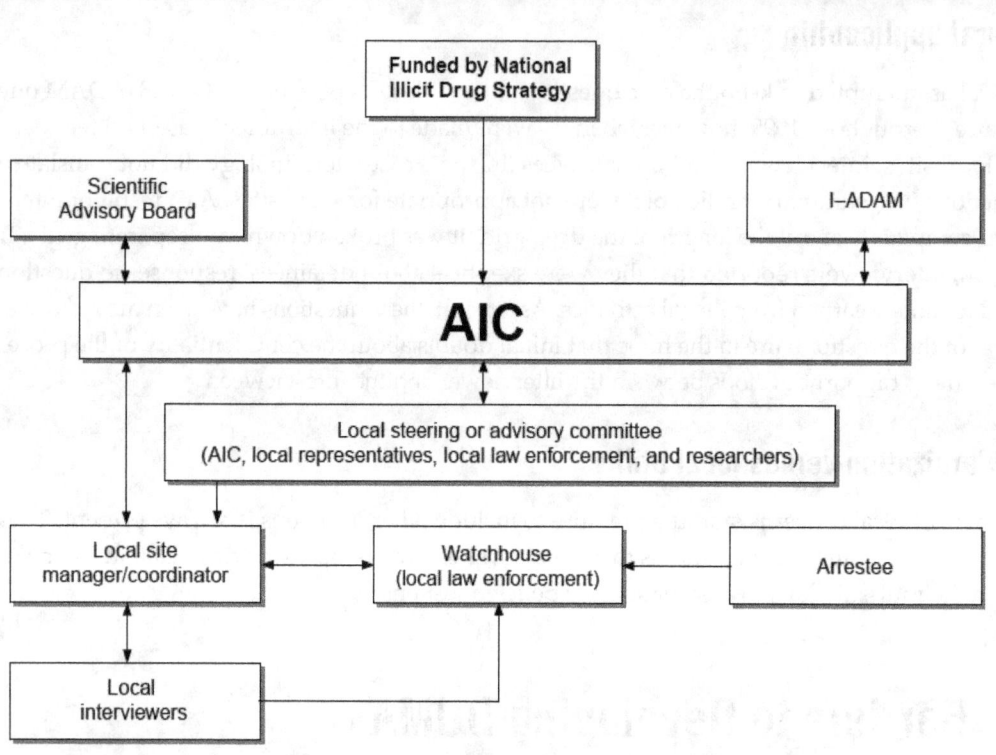

Each site has a local steering or advisory committee that supports local data collectors, monitors the local progress of the study, suggests ways of improving the project, and ensures that information is disseminated to relevant local agencies. The committees are made up of a cross-section of people that includes representatives from local law enforcement and researchers.

IX. Research Methods

No attempt was made to sample within the available pool of detainees. The selected sites were regarded as having a high volume of detainees relative to other possible sites. Even these sites, however, processed considerably fewer detainees than comparable U.S. sites. Maintaining 24-hour coverage would not have significantly increased the number of detainees interviewed and would have been prohibitively expensive. Interviewers entered the sites when the number of detainees was expected to be highest.

The major eligibility criterion was that the detainee had not been held in custody for more than 48 hours. Approximately 10 percent of detainees were deemed ineligible by local police, usually due to an assessment that there was a risk to the interviewer. Thus, the sample was not a random sample of all detainees brought to the police station or of all persons detained by the local police.

In all three jurisdictions—Queensland, Western Australia, and New South Wales—police are increasingly using notices to attend court or the equivalent that are issued on the spot rather than bringing people to the police station for questioning. Normally, these notices are for minor offenses when the person has no prior criminal record. DUMA has no access to these people.

Because the study is anonymous, interview subjects cannot be tracked across interview periods. Given that a substantial number of detainees self-reported that they had been arrested in the previous 12 months, it is likely that some detainees appeared in more than one quarter of interviews. Strictly speaking, the sample is one of detentions rather than detainees.

X. Training Issues

AIC has developed a training protocol manual for each site based on the U.S. National Institute of Justice protocols for the ADAM program and adapted to local conditions. Sites are required to train staff before each data collection. In addition, AIC produces a report for each site that details question-by-question errors by the interviewers to assist in the training process.

XI. National Data Analytic Results

The data represent four quarters of interviews for the Queensland and Western Australian sites and two quarters for the two New South Wales sites.[13] As a result, the total sample size for 1999 was much smaller for the New South Wales sites than for the other two sites (see exhibit 8). In 2000, each site was monitored every quarter. Because fewer women than men are detained by the police, the sample size for women was considerably smaller, which should be borne in mind when examining the data for females.

In 1999, the program interviewed 1,402 detainees, 1,366 of whom were defined as adults in their relevant jurisdiction. Urine specimens were collected from 978 adult and juvenile detainees. Both the interview and the provision of a urine specimen were voluntary; detainees could choose to complete the interview and not provide a specimen. Approximately 70 percent of both men and women who agreed to an interview also provided a urine sample. Only data from adults are presented in this section.

Exhibit 8. Gender Profile of Adult Detainees (%)

	Queensland	Western Australia	New South Wales (Bankstown)	New South Wales (Parramatta)
Sample size*	544	479	157	183
Females	15	12	18	14
Males	85	88	82	86

*Gender was not recorded for one person in Western Australia and two people in Bankstown, New South Wales.

Source: Australian Institute of Criminology, DUMA collection 1999 (computer file).

Across all four sites, detainees of both sexes were usually younger than age 30. The largest concentration by age tended to be in the 21- to 25-year-old group, as shown in exhibit 9.

Detainees were asked if they had been arrested (excluding their current arrest) and if they had spent time in prison in the previous 12 months. These data are presented in exhibit 10. The pattern is similar across the four sites: approximately half of detainees reported at least one prior arrest in the previous 12 months, and more than 15 percent had spent time in prison during the previous 12 months. Parramatta had a slightly higher rate of prior imprisonment for males than the other sites, and it had the highest percentage of detainees self-reporting a prior arrest in the previous 12 months. These data indicate that many detainees are cycling through the system and that interventions are needed to break this cycle.

During the DUMA pilot project, all detainees were interviewed regardless of the charge. Exhibit 11 shows what was determined to be the most serious charge at arrest for the detainees interviewed. The most serious charge was not always obvious from the charge sheet; interviewers recorded up to three charges. Information from all three charges was considered when determining the most serious charge: Violent offenses were assigned the most serious charge, followed by property, then drug, then traffic (including drunk driving) offenses. All remaining offenses were classified in the "other" category.

The charge profiles varied among the sites. Southport had the lowest percentage of adult male detainees charged with a violent offense and the highest proportion charged for traffic offenses (many for drunk driving). Almost half of the adult male detainees in Parramatta were charged with property offenses, as were approximately three-quarters of adult female detainees. Drug arrests across the four sites varied for adult males from 7 percent in Parramatta to 15 percent in Bankstown. East Perth had the highest percentage of "other" charges, many for outstanding warrants.

Exhibit 9. Age of Adult Detainees

	Sample size*	Age (years) (%)				
		17/18†–20	21–25	26–30	31–35	36+
Males						
Queensland	464	17	27	23	13	19
Western Australia	422	23	26	20	13	18
New South Wales (Bankstown)	129	22	30	22	12	15
New South Wales (Parramatta)	158	14	26	23	13	24
Females						
Queensland	80	25	31	20	5	19
Western Australia	56	20	32	21	7	20
New South Wales (Bankstown)	28	18	32	18	14	18
New South Wales (Parramatta)	25	20	36	24	8	12

*Sample size adds up only to 1,363 because gender was not recorded for one person in Western Australia and two people in Bankstown, New South Wales.

†Age of adult is 17 years in Queensland and 18 years in Western Australia; Bankstown, New South Wales; and Parramatta, New South Wales.

Source: Australian Institute of Criminology, DUMA collection 1999 (computer file).

Exhibit 10. Prior Arrest or Imprisonment of Adult Detainees

	Sample size	Arrested in previous year (%)	Sample size	Imprisoned in previous year (%)
Males				
Queensland	464	44	464	15
Western Australia	421	53	420	16
New South Wales (Bankstown)	126	49	128	14
New South Wales (Parramatta)	155	54	155	26
Females				
Queensland	80	39	79	19
Western Australia	56	64	56	21
New South Wales (Bankstown)	27	63	28	14
New South Wales (Parramatta)	25	56	24	14

Source: Australian Institute of Criminology, DUMA collection 1999 (computer file).

Exhibit 11. Most Serious Charge at Arrest (%)

	Violent offense	Property offense	Drug offense	Traffic offense	Other offense
Queensland					
Male (*n* = 459)	14	21	10	37	18
Female (*n* = 80)	8	32	15	33	13
Western Australia					
Male (*n* = 416)	19	22	9	14	36
Female (*n* = 55)	16	40	9	7	27
New South Wales (Bankstown)					
Male (*n* = 120)	21	26	15	19	19
Female (*n* = 27)	11	30	22	26	11
New South Wales (Parramatta)					
Male (*n* = 150)	17	47	7	13	16
Female (*n* = 25)	12	72	4	8	4

Source: Australian Institute of Criminology, DUMA collection 1999 (computer file).

Exhibit 12 shows the percentage of detainees who tested positive for drugs by gender and site. Drug use among detainees was common among the sites; approximately three-quarters of adult male detainees across all sites tested positive for any drug. There were, however, significant variations among the sites for individual drugs. A higher percentage of both male and female detainees tested positive for cannabis in Southport and East Perth than in the two Sydney sites. By contrast, a much higher percentage of detainees in the two Sydney sites tested positive for opiates; the rates for adult males in Sydney were almost double those in East Perth and triple those in Southport. Virtually no detainees tested positive for cocaine, except in Bankstown, where 12 percent of adult females tested positive for the drug. Only 4 percent of adult males tested positive for amphetamines in Bankstown, compared with between 12 and 14 percent in the other three sites. Multiple

Exhibit 12. Percentage of Adult Detainees Testing Positive for Any Drugs, by Sex

	Any drug*	Cannabis	Opiates	Cocaine	Amphetamines	Multiple drugs*
Males						
Queensland (n = 389)	73	65	13	1	12	18
Western Australia (n = 270)	74	61	22	<1	14	22
New South Wales (Bankstown) (n = 71)	78	52	45	3	4	27
New South Wales (Parramatta) (n = 91)	70	56	39	1	12	35
Females						
Queensland (n = 68)	65	59	25	2	10	28
Western Australia (n = 34)	77	59	44	0	27	47
New South Wales (Bankstown) (n = 17)	70	47	65	12	6	35
New South Wales (Parramatta) (n = 17)	71	47	59	0	6	35

*Any drug is defined as testing positive for cannabis, opiates, cocaine, or amphetamines. Multiple drugs is defined as testing positive for two or more of these drugs.

Source: Australian Institute of Criminology, DUMA collection 1999 (computer file).

drug use is common. It was highest overall among arrestees in Bankstown and Parramatta—the two sites with the highest rates of opiate use—although the single group with the highest prevalence rate was female arrestees in East Perth, 47 percent of whom used more than one drug.

Exhibit 13 shows that, consistently across all sites, adult male detainees tested positive for a range of drugs regardless of the charge for which they were detained. Males detained for minor offenses up to the most serious violent offenses tested positive for cannabis, opiates, and amphetamines. The positive test rates for cannabis were higher than for other drugs, in part because the test could detect cannabis use for up to 30 days, whereas it could detect use of opiates and amphetamines only within the previous 2 to 4 days.

Opiates were detected in similar and significant proportions across all offense categories in all sites except Southport. In Southport, 32 percent of property offenders tested positive for opiates, but among those charged with other offenses, relatively few (less than 12 percent) tested positive for opiates. In the other three sites, a significant percentage of detainees charged with a range of offenses tested positive for opiates; 23 percent of violent offenders in East Perth, 46 percent of those charged with "other" offenses in Bankstown, and 46 percent of those charged with traffic offenses in Parramatta. The strongest correlation in all sites was found between property crimes and the use of opiates. It was particularly strong in Bankstown, where 78 percent of property offenders tested positive for opiates.

As with opiate use, amphetamine use occurred across all offense categories. This applied less to the Sydney sites, which had a relatively small pool of detainees who tested positive for amphetamines. The correlation between violent offending and amphetamine use was nowhere near as

Exhibit 13. Percentage of Adult Male Detainees Testing Positive, by Most Serious Charge

	Sample size*	Cannabis (%)	Opiates (%)	Amphetamines (%)
Queensland				
Violent offense	57	56	9	11
Property offense	84	71	32	17
Drug offense	35	89	11	17
Traffic offense	143	64	6	12
Other offense	67	57	8	6
Western Australia				
Violent offense	52	60	23	15
Property offense	57	68	42	12
Drug offense	22	60	23	23
Traffic offense	38	63	16	16
Other offense	95	57	14	13
New South Wales (Bankstown)				
Violent offense	15	67	33	0
Property offense	18	50	78	6
Drug offense	9	67	33	11
Traffic offense	13	39	23	0
Other offense	13	46	46	8
New South Wales (Parramatta)				
Violent offense	13	54	23	15
Property offense	42	57	52	12
Drug offense	5	100	20	0
Traffic offense	13	31	46	15
Other offense	11	64	18	0

*Sample size equals number of adult male detainees who tested positive for any drug.

Source: Australian Institute of Criminology, DUMA collection 1999 (computer file).

strong as that found between opiates and property offending. In Bankstown, none of the violent offenders tested positive for amphetamines, but the sample size was small (15 offenders charged with a violent crime). At the other three sites, from 11 to 15 percent of those charged with violent offenses tested positive for amphetamines.

XII. Policy Implications and Applications of Data

A key goal of policing is to reduce crime. Given that certain criminal activities are closely associated with illicit drug use, monitoring the use of drugs by detainees is of key importance to law enforcement. DUMA provides for the first time a reasonable and independent indicator of drug-related crime in specific areas in Australia.

Australian law enforcement has not had a systematic monitoring system to track drug use among people who come into contact with criminal justice agencies. Much of the discussion on the links between drugs and crime has been based on anecdotal evidence or localized studies. More rigorous national collections are required for evidence-based policymaking. DUMA's goal is to overcome a significant limitation in Australia's national surveillance of illicit drug use by undertaking ongoing monitoring of detainees' involvement in drugs and crime.

Law enforcement concerns itself with both the demand for and the supply of illicit drugs. To enact successful policies to intervene in illicit drug markets, long-term monitoring of these markets is required. As with all commercial markets for major products, local markets are inextricably tied to global markets; one cannot be understood without the other. To understand supply, one must understand where, how, and when demand occurs and changes.

The purpose of DUMA is to improve understanding of the supply and demand for illicit drugs among detainees at the local level while providing comparable data across sites to facilitate the aggregation of national data. DUMA represents a research platform in the criminal justice system that can enable monitoring of supply and demand for illicit drugs at the local, national, and international levels.

DUMA data are now being used in three ways. First, local watchhouse personnel use the data to ensure a high standard of care for detainees. Appreciation of the types and amounts of drugs used by those coming into the watchhouse helps watchhouse staff better prepare for any medical situations they might face. Second, strategic law enforcement analysts at the State and local levels use the data to improve their understanding of drug consumption patterns and related criminal activities. Finally, local health agencies use the data to plan new drug treatment options.

Notes

1. Australia's literacy rate compares favorably with those of the United States and the United Kingdom. According to the International Adult Literacy Survey, 17 percent of Australian adults ages 16–65 are at the lowest level of literacy, compared with 23 percent in the United Kingdom and 24 percent in the United States. In contrast, 17 percent of Australian adults ages 16–65 are at the highest level of literacy, compared with 19 percent in both the United Kingdom and the United States. Organisation for Economic Cooperation and Development, *Education at a Glance 2000,* Paris: Organisation for Economic Cooperation and Development, 2000: table A2.3a.

2. Makkai, Toni, *Linking Drugs and Criminal Activity: Developing an Integrated Monitoring System,* Trends and Issues in Crime and Criminal Justice Series No. 109, Canberra: Australian Institute of Criminology, 1999.

3. Purity levels are calculated only on those seizures that have been analyzed at a forensic laboratory. See Australian Bureau of Criminal Intelligence, *Australian Illicit Drug Report 1998–99,* Canberra: Australian Bureau of Criminal Intelligence, 2000: tables 11.11, 11.13, and 11.14.

4. Ibid., tables 11.11, 11.13, and 11.19.

5. Australian Bureau of Statistics, *National Crime Statistics/Recorded Crime, Australia, 1993–2000*, Canberra: Australian Bureau of Statistics, 1994–2001, ABS 4510.0.

6. Wimshurst, Kerry, and Arch Harrison, "The Australian Criminal Justice System: Issues and Prospects," in *Crime and the Criminal Justice System in Australia: 2000 and Beyond*, ed. Duncan Chappell and Paul Wilson, Sydney: Butterworths, 2000.

7. For more detailed information, see Makkai, Toni, "Crime in Australia," in *Sociology of Australian Society*, ed. Jake Najman and John Western, Melbourne: Macmillan Publishers, 2000.

8. Makkai, Toni, "Harm Reduction in Australia: Politics, Policy, and Public Opinion," in *Harm Reduction: National and International Perspectives*, ed. James A. Inciardi and Lana D. Harrison, Thousand Oaks, California: Sage Publications, 2000: 171–192.

9. Ministerial Council on Drug Strategy, *National Drug Strategic Plan 1993–1997*, Canberra: Australian Government Printing Service, 1993: 4.

10. A more detailed outline of DUMA is provided in Makkai, Toni, *Drug Use Monitoring in Australia (DUMA): A Brief Description*, Research and Public Policy Series No. 21, Canberra: Australian Institute of Criminology, 1999.

11. More detailed information on urinalysis testing is provided in Australian Institute of Criminology, *DUMA: Drug Detection Testing*, Research in Public Policy Series No. 25, Canberra: Australian Institute of Criminology, 2000.

12. 42 U.S. Code 3789g and 28 Code of Federal Regulations Part 22.

13. This section of the chapter draws from data presented in Australian Institute of Criminology, *DUMA: 1999 Annual Report*, Research in Public Policy Series No. 26, Canberra: Australian Institute of Criminology, 2000.

CHILE

Luis Caris
Bruce Taylor

Luis Caris is a psychiatrist and professor at the School of Public Health, University of Chile, where he directs the drug prevention program. He also is chief of the Alcohol, Tobacco, and Drug Unit, Ministry of Health; a short-term consultant to the Pan American Health Organization; and a consultant to the Organization of American States–CICAD. **Bruce Taylor** is the deputy director of the Arrestee Drug Abuse Monitoring (ADAM) Program at the National Institute of Justice (NIJ).

I. Overview of Demographic Profile of Nation

Chile is a long, thin country in the extreme southwest of South America. Its closest neighbors are Argentina to the east and Bolivia and Peru to the north. Chile is more than 2,650 miles long but only about 300 miles wide at its widest point. Chile is as geographically isolated as it is diverse. To the south, it reaches into the Antarctic zone; to the west, the Pacific Ocean; to the north, the Atacama Desert; and to the east, the Andes.

Chile exhibits many of the traits that characterize Latin American countries. Its colonization by Spain began in the 16th century. The culture that evolved was largely Spanish; the influence of the aborigines is negligible. The country has a relatively homogenous, primarily mestizo population with a strong cultural unity (see exhibit 1 for demographic information). Chile has no sizable minority groups. During its history, a small elite has controlled most of the land, wealth, and political life. Chile did not depend as heavily on agriculture and mining as other Latin American countries; it also developed a manufacturing economy. Chile is one of the more urbanized Latin American countries, with a substantial middle class. More than 85 percent of the total population is urban; the metropolitan area of Santiago, the capital, has a population of more than 5 million.

The Republic of Chile was established in 1821. It is divided politically into 13 regions (Regions I–XII and the Santiago Municipal Region), 51 provinces, and 300 municipalities. Chile has a long history of representative democratic government and political freedom, with only a few short-lived exceptions. From September 1973 to March 1990, a military junta headed by General Augusto Pinochet Ugarte presided over the longest period of authoritarian rule in the country's history. Chile is governed in accordance with the Constitution of 1980,[1] approved by a plebiscite called by General Pinochet to change the Constitution of 1925. The 1980 Constitution places state administration in the hands of the President, who appoints the state ministers. The regions are headed by an intendente (intendant), provinces are headed by a gobernador (governor), and municipalities are headed by an alcalde (mayor). In towns with fewer than 10,000 inhabitants, the mayor is appointed by the municipal council; in towns and cities with more than 10,000 inhabitants, the mayor is appointed by the President of the Republic.[2]

Chile is one of Latin America's most economically developed countries, with a diversified free-market economy characterized by a high level of foreign trade (see exhibit 2). According to the human development index, Chile leads all Latin American countries in terms of development and is categorized as having high development in a world context, ranking 38th overall.[3]

Although Chile's crime rates in the early 1990s remained far below those of the United States, the notion that common crime was rare in Chile until the 1970s and 1980s is a myth. Chile's rural areas were plagued with outlaw gangs in the early 19th century. Attempting to analyze the crime situation under the military regime in the 1970s and 1980s is confusing because the government and media tended to lump ordinary criminal behavior together with dissident violence. Figures on crime and criminals released by the Instituto Nacional de Estadísticas (National Statistics Institute—INE) are not enlightening as to the true prevalence of crime in the country. Neither juvenile nor adult crime rates are broken down by gender, and no statistics are provided on prison populations. Few, if any, published sources provide general crime statistics,

Exhibit 1. General Demographic Information for Chile

Geographic area	756,950 km^2 (+ 1,249,675 km^2 Chilean Antarctic Territory*)
Estimated population	15,153,797
Population by ethnic origin	95% white/white-Amerindian, 3% Amerindian, 2% other
Religious affiliation	89% Roman Catholic, 11% Protestant, <1% Jewish
Gender distribution	50.5% female, 49.5% male
Age distribution	0–14 years, 28%; 15–64 years, 65%; ≥65 years, 7%
Life expectancy (at birth for 2000 cohort)	Average, 75.74 years; male, 72.43 years; female, 79.22 years
Population growth rate	1.17%
Birth rate	17.19 per 1,000
Total fertility rate	2.2 children per woman
Death rate	5.52 per 1,000
Literacy rate (age 15 and older)	Total, 95.2%; male, 95.4%; female, 95.0%
Language	Spanish
Urbanization rate	85.4%
Political subdivisions	13 regions (I–XII and Santiago Municipal Region), 51 provinces, 300 municipalities

Note: Demographic information is as of 2000, except for the literacy rate, which is estimated as of 1995.

*The United States does not recognize Chile's claim to the Chilean Antarctic Territory. Chile's territorial claim in Antarctica partially overlaps British and Argentine claims.

Sources: Central Intelligence Agency, *The World Factbook—Chile*, Washington, DC: Central Intelligence Agency, 2000, retrieved August 15, 2001, from the World Wide Web; http://www.odci.gov/cia/publications/factbook/geos/ci.html; Hudson, Rex A., *Chile—A Country Study*, Washington, DC: Library of Congress, Federal Research Division, 1994, retrieved October 24, 2000, from the World Wide Web: http://lcweb.loc.gov/frd/cs/cltoc.html.

with the main exception of figures for arrests, which are readily available in INE's *Compendio Estadístico*.[4] Chile's reported crime rate of about 1,300 crimes per 100,000 people is low compared with the reported U.S. rate of about 4,615 crimes per 100,000 people.[5]

Chile long remained relatively unaffected by either drug trafficking or extensive drug abuse. Small increases in both occurred during the late 1960s and early 1970s, reflecting an international trend.[6] By the early 1970s, Chile had become an important regional center for cocaine processing. Drug trafficking has been growing in Chile since the early 1980s, not only because of its geographic configuration and location bordering on the world s two leading producers of coca—Peru and Bolivia—but also because of its economic stability. With its open-market economy and bank secrecy laws, Chile is an attractive haven for money laundering. By 1992, Chile's most serious drug-related problem involved illicit drug transit along the northern corridor to Africa. In early 1993, a new cocaine/cocaine paste drug route reportedly came from Bolivia through the Azapa Valley.

After 1989, drug-related crime increased dramatically, particularly in the north of the country. In 1990, the police estimated that 20 percent of Africa's population aged 15 to 34 were habitual drug users. Of 385 homicides (0.3 per 10,000) in Chile during 1990, nearly 77 (20 percent) were classified as drug related.

Exhibit 2. Economic Information for Chile

Gross domestic product (GDP) (1999 est.)	$185.1 billion
GDP growth rate (1999 est.)	–1%
GDP per capita (1999 est.)	$12,400
GDP composition by sector (1999 est.)	Services, 61%; industry, 33%; agriculture 6%
Major export partners (1998)	European Union, 27%; United States, 16% Japan, 14%; Brazil, 6%; Argentina, 5%
Major export commodities	Copper, fish, fruits, paper, pulp, and chemicals
Labor force (1999 est.)	5.8 million
Labor force composition by sector (1997 est.)	Services, 59%; industry, 27%; agriculture, 14%
Poverty rate (1998 est.)	22%
Unemployment rate (1999 est.)	9%
Human development index (1998)	0.836

Note: The human development index measures a country's average achievement in three basic dimensions of human development: a long and healthy life, knowledge, and a decent standard of living. The index is a composite index of three variables: life expectancy at birth; educational attainment (adult literacy and the combined primary, secondary, and tertiary enrollment ratio); and per capita Gross Domestic Product (measured as purchasing power parity in U.S. dollars). Countries are grouped into three categories: high human development (≥0.800 on a scale of 0–1.000), medium human development (0.500–0.799), and low human development (<0.500) (United Nations Development Programme, *Human Development Report 2000*, New York: Oxford University Press, 2000, retrieved July 13, 2001, from the World Wide Web: http://www.undp.org/hdr2000).

Sources: Central Intelligence Agency, *The World Factbook—Chile*, Washington, DC: Central Intelligence Agency, 2000, retrieved August 15, 2001, from the World Wide Web: http://www.odci.gov/cia/publications/factbook/geos/ci.html; United Nations Development Programme, *Human Development Report 2000*, New York: Oxford University Press, 2000, retrieved July 13, 2001, from the World Wide Web: http://www.undp.org/hdr2000.

II. Description of the Criminal Justice System, Arrest Process, and Detention System

Criminal Code

The Criminal Code of Chile, first drafted in 1870 after two unsuccessful attempts, was promulgated in 1874 and modified in 1928 and again in 1930. Its models were the criminal codes of Austria, Belgium, France, and Spain. The death penalty, abolished in 1930, was reinstated for certain crimes in 1937 and abolished again in 2001.

The Criminal Code is divided into general and special sections. The general section enumerates general principles of criminal law relating to jurisdiction, the concept of crime, attempted crime, second-party participation in the commission of crime, habitual criminals, penalties, and circumstances that exclude or extinguish criminal responsibility. The special section defines specific offenses and their appropriate penalties. The courts are charged with ensuring that the penalty is appropriate not only to the crime but also to the criminal's culpability.[7]

Classification of Crimes

Crimes are divided into three basic categories: crímenes (serious crimes), delitos (minor crimes), and faltas (misdemeanors). A crime is defined as a voluntary act or omission for which the law imposes punishment. Criminal responsibility is specifically excluded in cases in which the defendant is insane or younger than 10. Minors 10 to 16 years old are not held criminally responsible unless it can be proven that they acted with full understanding of their acts. Criminal responsibility is also excluded for violent acts committed in the defense of one's own person, property, or rights; those of one's spouse; or those of a third party. Other excluded violent acts are those committed—

- Accidentally in the exercise of a legal act.
- In the exercise of public duty.
- Under duress or fear.
- Resulting in the killing or wounding of the accomplice of an adulterous spouse.

Criminal responsibility is also excluded for crimes of omission owing to a legal or irresistible cause. Suicide and attempted suicide are specifically decriminalized.[8]

Arrest Procedure

The Chilean Constitution states that no one may be arrested or detained unless on the order of a public official empowered by law. The police may detain someone caught in the act of committing a crime or based on suspicion. On detention, an arrestee must be brought before a competent judge within 24 hours. The Constitution allows civilian and military courts to order detention for up to 5 days without arraignment and to extend the detention of suspected terrorists for up to 10 days. The law affords detainees 30 minutes of immediate and subsequent daily access to a lawyer and to a doctor to verify their physical condition. The law does not permit a judge to deny such access. The Constitution allows judges to set bail.[9] Trial by jury is not provided for, and heavy reliance is placed on police evidence in criminal cases.

The Courts

The founders of Chile drew from the example of the United States in designing the institutions of government, and from Roman law and Spanish and French traditions, particularly the Napoleonic Code, in designing the judicial system. The 1925 Constitution introduced reforms aimed at depoliticizing and improving the judicial system by guaranteeing judicial independence. The organization and jurisdiction of Chile's courts were established in the Organic Code of the Tribunals (Law 7,241), adopted in 1943. This law has been modified on several occasions, mostly recently by Organic Constitutional Law 18,969 of March 10, 1990, and Law 19,124 of February 2, 1992. Chile's ordinary courts consist of juzgados de letras (local courts), major claims courts, the cortes de apelación (appellate courts), and the Supreme Court. There is also a series of special courts, such as the juvenile courts, labor courts, and military courts in peacetime. The local courts consist of one or more tribunals specifically assigned to each of the country's communes, Chile's smallest administrative units. In larger jurisdictions, the local courts may specialize in criminal or civil cases, as defined by law. Chile's 16 appellate courts each has jurisdiction over 1 or more provinces.[10]

The Penal System

Chile's penal system has been standardized since 1930, coming under the jurisdiction of the Minister of Justice. The system emphasizes rehabilitation of the offender. The degree of confinement is reduced progressively throughout the prisoner's sentence and ends, subject to good behavior, in conditional release for periods up to 50 percent of the total sentence. Chile has approximately 140 penal institutions with a total capacity of approximately 15,000 inmates; in 1990, however, the estimated number of prisoners was larger than 25,000. During the third quarter of 1990, for example, Santiago's San Bernardo Prison, designed to hold 800 prisoners, housed 3,300 inmates.[11]

III. Status of National Drug Policy

In 1999, a comprehensive revision of Chile's 1995 counternarcotics law was undertaken and proposed strengthening changes were drafted. The changes focus on authorizing flexibility in sentencing (currently, the minimum sentence is 5 years regardless of severity of the offense), combating money laundering through mandatory reporting of transactions and creation of a financial intelligence unit, and harmonizing the drug code with Chile's ongoing legal reform.[12]

The Chilean Government's coordinating body for drug policy and demand reduction activities is the Consejo Nacional para el Control de Estupefacientes (National Drug Control Council—CONACE). CONACE's 2001 annual budget of approximately $15.2 million represents a real increase of more than 30 percent from its 2000 budget. Of CONACE's 2001 budget, $8.2 million is earmarked for demand reduction programs—$3.1 million for community-based prevention programs, $2.6 million for prevention targeted at schools, and $2.5 million for drug treatment. Resources to be channeled to school- and community-based programs will double under the 2002 budget.[13]

In a major drug policy initiative in 1998, CONACE decentralized its national project bank for drug prevention, education, and rehabilitation. As a result of the decentralization, regional councils and local municipalities gained more input in projects that account for approximately 90 percent of the national project bank's funds. Individual projects also can be proposed and implemented by nongovernmental organizations, local governments, universities, and other institutions.[14]

Chile is not considered a center of production of illegal drugs. Partly because of its geographic isolation, it has until recently avoided many of the problems created by illegal narcotics in other South American countries.[15] As a consequence of the expansion of drug trafficking and narcotics abuse during the late 1960s and early 1970s, Chile's first antinarcotics law had been passed by 1973. Later, the Government formed a special narcotics unit and began a highly effective crackdown, bringing the narcotics problem under control within a year. Government antidrug forces also pioneered a youth education program against narcotics. The dismantling of a large drug-trafficking and money-laundering ring in May 1998 demonstrates the Government's continued efforts to counter drug smuggling and related criminal activity.[16]

Chile is a party to the 1988 United Nations Drug Convention. Government and private organizations support drug-abuse prevention and rehabilitation programs. Chile continues to play a leading role in the Organization of American States' Comisión Interamericana para el Control del Abuso de Drogas (Inter-American Drug Abuse Control Commission—CICAD), including its efforts to create a multilateral evaluation mechanism to combat illegal drug trafficking and abuse. Chile worked closely with CICAD in the area of money laundering in 1998 and sponsored public- and private-sector seminars on the topic.[17] Chile has joined the Grupo de Acción Financiera de Sudamérica Contra el Lavado de Activos (South American Financial Action Task Force on Money Laundering—GAFISUD), created in December 2000. In addition, in August 2000, the U.S. and Chile signed an umbrella agreement to share information, strategies, and resources to continue joint efforts against narcotics.[18]

Chile has several ongoing data collection systems that help inform drug policy. I—ADAM will serve as another indicator of drug use in Chile. Currently, the Chilean Government conducts a door-to-door household drug survey of a representative sample of the population every 2 years, a school-based drug survey, a study of emergency rooms similar to the Drug Abuse Warning Network program in the United States,[19] and various treatment facility surveys.

IV. Background History of the I–ADAM Program in Chile

In 1994, the Ministry of Health set up an alcohol and drug surveillance/monitoring system aimed at assessing alcohol and drug abuse in Chile. Monitoring was conducted in several cities and data were gathered on drug type, drug and alcohol abuse, and health outcomes. Public school students and people entering hospital emergency rooms, jail detention centers, and treatment centers answered self-administered questionnaires about the health risks associated with alcohol and/or drug use, protective behaviors, prevention, and consumption. Data were analyzed to assess the magnitude of alcohol and drug abuse, and an early warning system against the use of alcohol and/or drugs was set up. As previously mentioned above, the Government conducts a door-to-door drug survey of a representative sample of the population every 2 years. Although research has been done in Chile on the relationship between health and drugs/alcohol, the relationship between drugs and crime has yet to be explored. In 1998, Chile implemented an I–ADAM project to explore this relationship and to serve as an indicator of drug use in the country.

As a precursor to its participation in the I–ADAM project, Chilean health authorities conducted a 10-day study of drug abuse among detained arrestees in March 1997. Five hundred forty-nine arrestees (499 men and 50 women) from six police stations in the Santiago municipal region participated in the research. The police stations were located in Santiago city, Renca, Nuñoa, Peñalolén, San Miguel, and Conchalí. Chilean police bring all persons arrested for nontraffic offenses to the police stations for booking. During this 10-day period, researchers attempted to interview all arrestees: 75 percent of the arrestees approached by the research staff agreed to take part in the study, and 75 percent of those who participated agreed to provide a urine specimen. Because of budget restrictions, urine testing was done only for cocaine and marijuana. The study instrument and a report of this study are available only in Spanish.

V. Goals and Objectives of I–ADAM in Chile

The primary aim of I–ADAM in Chile is to generate drug abuse surveillance/monitoring data. Program objectives include the following:

- Testing the applicability of the I–ADAM research methodology.
- Assessing the applicability of the I–ADAM instrument.
- Describing the types and forms of drug use in detention centers.
- Generating information concerning the magnitude of substance abuse in the arrestee population.

VI. Getting Started

A. Obtaining Funding

The Ministerio de Salud (Ministry of Health) funds and operates the I–ADAM program in Chile. The Ministry of Health has five divisions, one of which is the División de Salud de las Personas (Division of the People's Health). This division has three components: the Departamento de Programas de Las Personas (People's Health Program), Departamento de Epidemiología (Department of Epidemiology), and Departamento Odontológico (Department of Odontology). I–ADAM in Chile is operated as a research program of the People's Health Program; its name is Vigilencia Epidemiologica Drogas en Chile (Drug Epidemiological Surveillance in Chile).

Because Chile's I–ADAM program is administered by employees of the Ministry of Health, nearly all the costs associated with this research are absorbed as part of employees' salaries. The only extra costs incurred by the I–ADAM program are for the laboratory costs of urine testing and overtime pay for the field site director and the team of interviewers to conduct the research interviews. I–ADAM management aimed for two data collection points per year at each site. Funds were allocated for one data collection point in 1998 and for two data collection points in 1999. Efforts are under way to seek funding from the Ministries of Health and Justice and from foreign and multinational agencies and foundations operating in Chile.

B. Contracting for Data Collection

As indicated above, this project was conducted in house by staff of the Ministry of Health. Hiring a contractor to conduct this work was considered cost prohibitive.

C. Facility Access

Access to the lockup facilities was secured through a partnership between the Ministry of Health and the Carabineros (the national uniformed police). The mayors (commanders) of the comisarías (police stations) gave research staff written permission to conduct the study and gain access to the jails. Because police were not accustomed to having research conducted in their jails, they were briefed in detail about the scope of the project.

D. Knowing the Catchment Area

The 1999 I–ADAM study was conducted in select comisarías in three municipalities of the capital city of Santiago: the First Police Station of Santiago, the Fifth Police Station of Conchalí in the northern part of the city, and Lo Espejo in the southern part of the city. Data collection sites were selected based on their case flow and neighborhood characteristics. One site is in the heart of a thriving downtown commercial area that has arrestees from all socioeconomic levels; another site is an active drug-trafficking area; the third site has many drug addicts. Future research sites will include police detention centers in either Valparaiso or Inquinque. Collaboration among the Ministry of Health, the director general of the Carabineros, and police station commanders made access to selected research sites easier.

E. Drug Testing Issues

Because of the relative scarcity of high-volume drug-testing equipment in Chile, drug testing is expensive in the country. After a search, it was determined that PharmChem, the U.S. laboratory that conducts the U.S. ADAM drug testing, offered the most affordable option for the I–ADAM group in Chile. PharmChem used the EMIT™ (enzyme multiplied immunoassay testing) system—the same system used by the U.S. ADAM program—to screen for the presence of drugs in urine. Samples were tested for alcohol, amphetamines, barbiturates, benzodiazepines, cocaine, marijuana, methadone, methaqualone, opiates, PCP, and propoxyphene.

F. Instrument Issues

The questionnaire used in the I–ADAM Chile study was based on the early Drug Use Forecasting (DUF) questionnaire.[20] The National Institute of Justice (NIJ) translated the DUF questionnaire and sent it to the Chilean research team for review. Minor edits and adaptation to the Chilean idiom were made, and it was deemed necessary to add as well as remove a few questions on the basis of cultural applicability. Several drug market questions were dropped, and the list of drugs to be queried on the drug grid was revised. The self-report survey includes the following drugs: alcohol, marijuana, coca paste, crack cocaine, powder cocaine, heroin/opiates/dilaudid, amphetamines, methamphetamines, inhalants, and tranquilizers. Some housing questions, illegal income questions, and the ethnicity question were removed.

Cultural applicability

Spanish is the official language of Chile, and virtually everyone is conversant in Spanish. Therefore, it was decided to use only one questionnaire designed in Spanish. Arrestees who could not speak Spanish were deemed ineligible and were excluded from the survey.

Standardization versus local utility

The same questionnaire is used across all sites in Chile. It was not found necessary to adjust the questions for local cultural differences. Because cultural standardization is particularly relevant when making comparisons across countries, the I–ADAM Chile survey includes a series of core questions that are used in the ADAM surveys in the United States and other I–ADAM participating countries.

VII. Barriers to Developing I–ADAM

To date, the development of the Vigilencia Epidemiologica Drogas en Chile has run smoothly. Nonetheless, barriers arose at the onset of the program. The primary barrier to establishing I–ADAM in Chile has been obtaining funding and establishing the legitimacy of the research. Full government support and funding is needed to implement the program. Although the Ministry of Health earmarked funds for the program in 1999, 2000, and 2001, long-term funding has not yet been obtained. As noted earlier, program managers are soliciting the Ministry of Health, the Ministry of Justice, and foreign sources of funds such as CICAD, the United Nations Drug Control Programme, and the American Embassy in Chile.

VIII. National Stakeholders Involved in the Development of I–ADAM

The key players involved in developing and implementing I–ADAM in Chile are the Ministry of Health (the minister and undersecretary), the Ministry of Justice (the Ministry's representative to CONACE, the acting undersecretary, and the division directors), and the Carabineros (the director of the Departamento Control Drogas y Prevención Delictual—Department of Drug Control and Crime Prevention).

A cooperative relationship has been established among representatives of these Government agencies. The Ministries of Health and Justice have expressed an interest in ADAM as a platform for developing policy-relevant research pertaining to drug control activities and the health consequences associated with drug use. The director of the Department of Drug Control and Crime Prevention has expressed a special interest in international research, especially the potential for comparative data applications.

IX. Research Methods

The I–ADAM program in Chile has two fundamental components: a questionnaire administered in a booking facility by a trained interviewer to an arrestee within 48 hours of the arrest and a urine specimen collected from the respondent that is used to corroborate statements about recent drug use.

Each of Santiago's 49 municipalities has its own police force. Chile has a uniformed police force (Carabineros) and a nonuniformed police force (Policia de Investigaciones) similar to the U.S. Federal Bureau of Investigation. Nonuniformed officers work at the federal level and deal with national issues. A judicial order must be issued before nonuniformed officers can initiate an investigation. Uniformed officers staff comisarías and make arrests. Only persons arrested by uniformed officers are eligible for inclusion in the I–ADAM study.

Anyone arrested for a crime, including the least serious nontraffic offense, is brought to the municipal police station for at least 1 hour. The arrest information is entered in a logbook and recorded electronically. All persons arrested are eligible to be interviewed by I–ADAM research staff. Using the police station's computer, a list of all cases for any time period can be easily reproduced and used for sampling.

Given that the 1999 I–ADAM Chile study was a pilot test, the main purpose of which was to resolve implementation issues, nonprobability-based sampling methods were used. The three sites were selected intentionally because of their locations and neighborhood characteristics; they do not represent the population of arrestees in the nation or in Santiago. As mentioned earlier, some effort was made to choose a varying group of catchment areas. At the site level, all cases that met the study eligibility criteria were interviewed. Interviews were done from 8 a.m. to 10 p.m.; no provision was made to compare cases that were not sampled from 10:01 p.m. to 7:59 a.m.

The eligibility criteria for I–ADAM in Chile are the same as in the other I–ADAM participating countries. Data were collected from adult booked/processed arrestees detained long enough to be interviewed (but less than 48 hours). Study participants generally were interviewed before they saw a magistrate or judge, but they had access to legal counsel to ask about the voluntary and confidential nature of the study. Arrestees who did not speak Spanish or who were deemed too intoxicated, violent, or mentally unstable did not participate. The 1999 Chile pilot study included only men.

X. Training Issues

The goal of the I–ADAM training in Chile is to convey to a team of interviewers the practices and procedures required to conduct scientific research surveys. The training curriculum for the Chilean I–ADAM data collection team was based largely on the U.S. ADAM training manual. The development of the curriculum was a joint effort between NIJ and the Chilean Ministry of Health. Each Chilean interviewer received training using a standard curriculum and manual that addressed basic and advanced interview skills and site operating procedures. Training was conducted 2 to 3 days before actual data collection. The training curriculum covered the roles and responsibilities of key site personnel, a question-by-question review of the survey, the handling and shipping of urine specimens, paired mock exercises, editing exercises (e.g., editing the survey instrument for proper skip-pattern usage during the interview), interviewer safety, and a review of the concept of informed consent. Interviewers were required to complete and demonstrate competence in basic interviewing skills and an understanding of project operating procedures.

NIJ staff met with staff from the Ministry of Health in November 1998 to visit the facilities, develop a data collection plan, and meet with stakeholders from the Ministries of Health and Justice. In January 1999, shortly before data collection started, a U.S. training team was sent to Chile to help implement the first training of the Chilean field staff. Nurses from the Ministry of Health conducted the I–ADAM interviews.

The group of nurses selected for I–ADAM Chile had extensive experience with arrestees, having been involved in an earlier study of drug use among detainees. Using nurses to conduct interviews had advantages and disadvantages. Arrestees are reluctant to trust the police and to participate in a research project that looks as if it is sponsored by the police. Using nurses clearly shows the arrestees that the program is sponsored by the Ministry of Health and the information will not be used against them. The Ministry of Health is seen as nonrepressive and has a positive image as a healing organization. One belief was that if the research team did not use uniformed nurses, few arrestees would participate and give honest responses. The disadvantage to using nurses, especially in uniforms, is that some comparability problems will emerge with other countries. That disadvantage, however, is offset by the need to collect valid local data.

XI. Data Analytic Results

The initial I–ADAM study in Chile took place January 22–28, 1999, in three detention centers in the capital city of Santiago: the First Police Station of Santiago (First of Santiago), the Fifth Police Station of Conchalí, and Lo Espejo.

Uniformed nurses employed by the Ministry of Health conducted the interviews. Every eligible male arrestee, regardless of the offense with which he was charged, was interviewed. Interview hours were from 8 a.m. to 10 p.m. Interviews were conducted within 6 hours of the arrestees' arrival at the detention center and lasted an average of 10 minutes. The interview process went smoothly; the police took no action that could have influenced the number of respondents or the quality of the data collected.

Sample Characteristics

Interviews were conducted with 134 adult males in three locations. The largest number ($n = 99$) of respondents interviewed were brought to the First of Santiago detention center. When asked, 92 (68.7 percent) of 134 respondents provided urine samples. Two samples were too small to be analyzed, however, so only 90 samples were tested.

The age range of the study respondents was from 12 to older than 45; the average age was 29. Comparisons of the detention centers show that 12- to 19-year-olds made up more than 23 percent of the sample in Conchalí, but 30- to 44-year-olds made up more than 71 percent of those interviewed at Lo Espejo and about 50 percent of those interviewed at First of Santiago. As seen in exhibit 3, most arrestees interviewed (50.8 percent) were single. Most interviewees (61.3 percent) had a full- or part-time job.

Only 1.5 percent of arrestees reported having no formal education; 48.1 percent had attended high school. Although none of the respondents at either First of Santiago or Lo Espejo reported receiving assistance from government programs, 6.3 percent of those interviewed at the Conchalí detention center were on public assistance. Exhibit 3 shows that the largest group of respondents were arrested for drug or alcohol offenses in Conchalí and Lo Espejo. In First of Santiago, the largest group of respondents were arrested for illegal street vending.

Exhibit 3. Sociodemographic Profile of Respondents in Detention Centers

	First of Santiago (*n* = 99)	Conchalí (*n* = 21)	Lo Espejo (*n* = 14)	Total (*n* = 134)
Age (%)				
12–19	11.1	23.8	6.1	11.9
20–29	33.3	28.6	21.4	31.3
30–44	49.5	38.1	71.4	50.0
45 and older	6.1	9.5	1.1	6.7
Marital status (%)				
Single	53.6	42.9	42.9	50.8
Married/living together	38.1	38.1	57.1	40.2
Separated/widowed	8.2	19.0	—*	9.1
Educational level (%)				
No education	1.0	—	7.7	1.5
Primary school	42.4	28.6	30.8	39.1
High school	42.4	66.7	61.5	48.1
University	14.1	4.8	—	11.3
Source of income (%)				
Government assistance	—	6.3	—	0.8
Full-time employment	15.6	25.0	53.8	21.0
Part-time employment	40.0	50.0	30.8	40.3
Other	44.4	18.8	15.4	37.8
Monthly income (pesos)				
Less than 100,000	23.3	18.8	42.9	25.0
100,000–149,999	33.7	18.8	14.3	29.3
150,000–199,999	23.3	43.8	7.1	24.1
200,000 and above	19.8	18.8	35.7	21.6
Offense type				
Violent crime	5.1	23.8	14.3	9.0
Drug and/or alcohol offense	18.2	52.4	64.3	28.4
Property crime	7.1	—	7.1	6.0
Other offense	20.2	23.8	14.3	20.1
Street vendor	49.5	—	—	36.6
Detained last year				
Yes	57.1	28.6	46.2	51.5
No	42.9	71.4	53.8	48.5

*— = Not applicable.

Urine Test Results

Exhibit 4 shows the test results from the urine samples taken from interviewees. Approximately 48 percent of the respondents tested positive for at least one illicit drug (excluding alcohol), and 12.2 percent tested positive for two or more illicit drugs (9 of the 11 multiple drug users tested positive for cocaine and marijuana). The highest drug prevalence rate was for marijuana, followed by cocaine, alcohol, benzodiazepines, and amphetamines. No prevalence was detected for the six other drugs tested (barbiturates, propoxyphene, methaqualone, PCP, opiates, and methadone).

Respondents' Self-Reported Drug Use

Exhibit 5 shows respondents' self-reported drug use during the 3 months before the I–ADAM interview. Forty-one percent reported having used an illegal drug other than alcohol; 66.4 percent reported having consumed alcohol. Rates of use of illegal drugs ranged from 32.8 percent for marijuana to 0.7 percent for inhalants. No interviewees reported having used amphetamines or methamphetamines in the previous 3 months, although one interviewee tested positive for amphetamines (see exhibit 4).

Exhibit 6 shows information gathered about arrestees who used all drugs, including coca paste, powder cocaine, or cocaine in any other form. Information gathered includes marital status, age, education levels, rates of treatment, and rates of multidrug use. Cocaine users also reported significant levels of use of marijuana, amphetamines, and tranquilizers. This was the case even

Exhibit 4. Urine Test Results for I–ADAM Interviewees ($n = 90$)

Substance found	Prevalence (%)
Marijuana ($n = 28$)	31.1
Cocaine ($n = 24$)	26.7
Alcohol ($n = 18$)	20.0
Benzodiazepines ($n = 2$)	2.2
Amphetamines ($n = 1$)	1.1
Barbiturates	0.0
Methadone	0.0
Methaqualone	0.0
Opiates	0.0
PCP	0.0
Propoxyphene	0.0
Any of 9 drugs (excluding marijuana and alcohol) ($n = 25$)	27.8
Any of 9 drugs (excluding marijuana but including alcohol) ($n = 39$)	43.3
Any of 10 drugs (including marijuana but excluding alcohol) ($n = 43$)	47.8
Multiple drugs for 10 drugs (including marijuana and excluding alcohol) ($n = 11$)	12.2
Any of 10 drugs (including marijuana and alcohol) ($n = 54$)	60.0
No drugs ($n = 36$)	40.0

though no interviewees reported having used amphetamines in the previous 3 months (see exhibit 5). Exhibit 7 shows the reported mean age of first use of certain drugs, ranging from age 15 for inhalants to age 24 for coca paste.

XII. Policy Implications and Applications of Data

The initial I–ADAM pilot data for Chile show a high level of drug and alcohol use (60 percent of participants tested positive for drugs and/or alcohol) among arrestees in the three Santiago sites. The data may overstate the level of drug use among all arrestees, however, given that two of the three pilot sites (Conchalí and Lo Espejo) were chosen in part because of the high levels of drug-related activity in those areas and that, in those two areas in particular, a majority of those interviewed had been arrested for drug-related offenses (see exhibit 3). As the number of interview sites increases and as the I–ADAM project moves into metropolitan areas outside Santiago, the interview samples will become more representative of the arrestee population as a whole, and comparisons of arrestee drug use among metropolitan areas can be made.

One unique factor of Chile's I–ADAM program is the use of uniformed nurses from the Ministry of Health to interview the arrestees. The rationale behind this has been to gain the confidence of the arrestees and to encourage participation in the project. Nonetheless, urine testing of arrestees found that, as in other countries, a higher proportion of arrestees tested positive for drugs (47.8 percent) than admitted to drug use within the previous 3 months (41.9 percent). This indicates that some portion of arrestees will not be forthcoming about current drug use, even if there is no incentive for them to conceal their drug use.

The involvement of the Ministry of Health in the I–ADAM program at a basic level may have another benefit, however, in fostering closer cooperation between the Ministry of Health and other I–ADAM stakeholders (the Carabineros, CONACE, and the Ministry of Justice) in the

Exhibit 5. Respondents' Self-Reported Drug Use

Self-reported drug use over the previous 3 months	Prevalence (%)
Any drug, excluding alcohol	41.0
Alcohol	66.4
Marijuana	32.8
Coca paste and powder cocaine	21.6
Coca paste	15.7
Powder cocaine	9.7
Stimulants	6.0
Tranquilizers	6.0
Inhalants	0.7
Amphetamines	0.0
Methamphetamines	0.0

Exhibit 6. Characteristics of Users of Specified Drugs

Characteristic	All drugs (%)	Coca paste (%)	Powder cocaine (%)	Any cocaine* (%)
Reporting use in past 3 months	41.0	15.7	9.7	21.6
Reporting use by site				
Lo Espejo	35.7	35.7	0	35.7
First of Santiago	45.5	15.2	12.1	23.2
Conchalí	23.8	4.8	4.8	4.8
Primary age cohort of drug users				
20–29 years	32.7	26.8	53.8	25.0
30–44 years	54.5	66.7	30.8	62.5
Highest educational level				
Primary school	37.0	50	38.5	47.8
High school	46.3	35	30.8	39.1
University	14.8	10	30.8	13.0
Drug users married/living with someone	46.3	55	61.5	47.8
Reporting use of other drugs				
Marijuana	41.8	0	0	26.1
Powder cocaine	25.5	9.5	—	—
Coca paste	10.9	—	9.5	—
Amphetamines	—	9.5	53.1	0
Tranquilizers	—	9.5	23.4	0
Receiving drug treatment	5.5	9.5	7.7	4.2
Arrested for street drug sales	39	42.9	38.5	29.2
Arrested for other drug/alcohol offenses	24.5	23.8	46.2	33.3

— = Not applicable.

*Coca paste and powder cocaine.

Exhibit 7. Mean Age of First Consumption of Drugs

Drug consumed	Reported age
Inhalants	15
Alcohol	16
Marijuana	17
Powder cocaine	22
Coca paste	24

development of drug policy in general. In particular, this may lead to greater cooperation among the stakeholders in making drug treatment and drug education programs offered by the Ministry of Health available to arrestees and to the prison population.

Notes

1. The current Constitution of Chile was approved by a plebiscite on September 11, 1980, and became effective March 11, 1981. The Constitution was again approved and amended by referendum on July 30, 1989, and amended in 1993. (*Constitution of the Republic of Chile,* Santiago: Government Junta, 1980, retrieved October 24, 2000, from the World Wide Web: http://www.richmond.edu/~jpjones/confinder/Chile.htm; Central Intelligence Agency, *The World Factbook—Chile,* Washington, DC: Central Intelligence Agency, 2000, retrieved August 15, 2001, from the World Wide Web: http://www.odci.gov/cia/publications/factbook/geos/ci.html; Hudson, Rex A., *Chile—A Country Study,* Washington, DC: Library of Congress, Federal Research Division, 1994, retrieved October 24, 2000, from the World Wide Web: http://lcweb2.loc.gov/frd/cs/cltoc.html.)

2. Hudson, *Chile—A Country Study* (see note 1).

3. United Nations Development Programme, *Human Development Report 2001,* New York, Oxford University Press, 2001, retrieved August 30, 2001, from the World Wide Web: http://www.undp.org/hdr2001.pdf.

4. Instituto Nacional de Estadísticas, *Compendio Estadístico 2000,* Santiago: Instituto Nacional de Estadísticas, Departamento de Servicios al Usuario y Difusión, 2000, retrieved August 15, 2001, from the World Wide Web: http://www.ine.cl/chileci/compen/compind.htm.

5. U.S. Census Bureau, *Statistical Abstract of the United States 2000: The National Data Book,* Washington, DC: U.S. Department of Commerce, Economics and Statistics Administration, U.S. Census Bureau, 2000, retrieved August 30, 2001, from the World Wide Web: http://www.census.gov/prod/www/statistical-abstract-us.html.

6. Hudson, *Chile—A Country Study* (see note 1).

7. Ibid.

8. Ibid.

9. Bureau of Democracy, Human Rights, and Labor, *Human Rights Report, U.S. Department of State—Chile Country Report on Human Rights Practices for 1997,* Washington, DC: U.S. Department of State, 1998, retrieved August 15, 2001, from the World Wide Web: http://www.state.gov/www/global/human_rights/1997_hrp_report/chile.html.

10. Hudson, *Chile—A Country Study* (see note 1).

11. Ibid.

12. Bureau for International Narcotics and Law Enforcement Affairs, *International Narcotics Control Strategy Report, 2000—South America,* Washington, DC: U.S. Department of State, 2001, retrieved August 30, 2001, from the World Wide Web: http://www.state.gov/g/inl/rls/nrcpt/2000/index.cfm?docid=883.

13. Ibid.

14. Bureau for International Narcotics and Law Enforcement Affairs, *International Narcotics Control Strategy Report, 1998—South America,* Washington, DC: U.S. Department of State, 1999, retrieved August 30, 2001, from the World Wide Web: http://www.state.gov/www/global/narcotics_law/1998_narc_report/samer98_part2.html.

15. Ibid.

16. Ibid.

17. Ibid; Hudson, *Chile—A Country Study* (see note 1).

18. Bureau for International Narcotics and Law Enforcement Affairs, *International Narcotics Control Strategy Report, 2000—South America* (see note 12).

19. The Drug Abuse Warning Network (DAWN) is a large-scale, ongoing drug abuse data collection system sponsored by the Substance Abuse and Mental Health Services Administration of the U.S. Department of Health and Human Services. DAWN's major objectives are to identify substances associated with drug-related episodes and deaths reported by emergency departments and medical examiners, monitor trends in drug use consequences, detect abuse of new drugs, and assess health hazards associated with drug use. DAWN abstracts data from records of drug-related visits to emergency departments of acute care hospitals in the United States and drug-related deaths reported by 145 medical examiner jurisdictions in 43 metropolitan areas. Since 1988, DAWN emergency department data have been collected from a representative sample of 685 hospitals, including 21 oversampled metropolitan areas. DAWN medical examiner data are not nationally representative.

20. The Drug Use Forecasting (DUF) program was the predecessor to the ADAM program. Starting in eight U.S. cities in 1987, DUF was the first program to examine the prevalence of drug use among arrestees using a combination of interviews and drug testing of a sample of arrestees. For a history of DUF, see Reardon, Judy A., *Drug Use Forecasting Program: Measuring Drug Use in a Hidden Population,* Washington, DC: U.S. Department of Justice, Office of Justice Programs, National Institute of Justice, 1993, NCJ 144784.

ENGLAND AND WALES

Trevor Bennett

Trevor Bennett is the director of the New England and Wales Arrestee Drug Abuse Monitoring (NEW–ADAM) Program. He is a university lecturer at the Institute of Criminology, University of Cambridge, and a fellow of Wolfson College, Cambridge. He has served as acting director and deputy director of the Institute of Criminology.

I. Overview of Demographic Profile of Nation

The United Kingdom is a union of four countries: England, Wales, Scotland, and Northern Ireland. The New England and Wales Arrestee Drug Abuse Monitoring (NEW–ADAM) Program operates, as the name suggests, in England and Wales. Scotland is operating its own pilot program, and Northern Ireland has not yet begun to implement a program. NEW–ADAM is a national program that interviews arrestees and tests them for drugs. Surveys are conducted in 16 locations in 13 of the 43 police force areas in England and Wales. Fifteen of the 16 locations are in England; 1 is in Wales. In one force area, the Metropolitan Police District in London, four sites were selected. The survey sites are evenly spread geographically and cover a range of metropolitan and large urban areas as well as some smaller and rural districts.

The United Kingdom is a parliamentary democracy. The Parliament in London has full responsibility for England and Wales and partial responsibility for Scotland and Northern Ireland, which are developing their own national assemblies or parliaments as part of the national devolution of power. Local governments, which are accountable to the central government, decide how counties and cities are administered.

England and Wales have a combined population of approximately 51 million (see exhibit 1), of which nearly 25 million are men and slightly more than 26 million are women. About 25 percent of the population is younger than age 20. Ninety-four percent of the population is white. The major nonwhite ethnic groups are black (1.7 percent), Indian (1.6 percent), Pakistani and Bangladeshi (1.2 percent), and Chinese and other (1.2 percent).

Exhibit 2 shows crime trends over the past decade using data from the British Crime Survey (BCS), a national household survey of self-reported crime victimization, and officially recorded crimes submitted by the police to the Home Office. The BCS is based on a nationally representative sample

Exhibit 1. Demographic and Statistical Profile of England and Wales

Population size (1991)	51,100,000
Population change (%) (1901–91)	+57
Area (square miles)*	58,366
Age distribution (1991) (% < 20) / (% ≥ 20)	25.4 / 74.6
Gender distribution (1991) (% females) / (% males)	51.1 / 48.9
Ethnic distribution (% white)	94
Live birth rate (per 1,000 population)	13.7
Death rate (per 1,000 population)	11.2
Life expectancy (% of population age 85 and older)	1.6
Gross average household income (£)	£19,680[†]
Gross domestic product (1991)	£582,946m[‡]
Unemployment rate (1991) (% unemployed age ≥ 16)	8.4[§]
Urban living (% of land area designated as nonrural)	54
Poverty rate (% of households below 60% median income)	18[**]
Literacy rate (1996)	55

*Includes both land and water.

[†] Includes retired and nonretired households. Figure is for the financial year 1997–98 in the United Kingdom, including Scotland and Northern Ireland.

[‡] Gross domestic product at market prices: output approach. Covers the United Kingdom, including Scotland and Northern Ireland.

[§] Males and females. Figure includes Scotland and Northern Ireland. Rates calculated as a percentage of the economically active population.

**For Great Britain, including Scotland.

Sources: Wisniewski, D., ed., *Annual Abstract of Statistics: 2000 Edition*, London: Her Majesty's Stationery Office [HMSO], 2000; Matheson, J., and C. Summerfield, Social Trends 30: *2000 Edition*, London: HMSO, 2000; United Nations Development Programme, *Human Development Report 1996*, Oxford: Oxford University Press, 1996.

Exhibit 2. Crime Trends in England and Wales

	Number of crimes (in millions)								
	1991	1992	1993	1994	1995	1996	1997	1998*	1999
British Crime Survey (national household survey of victim-reported crime)[†]	15.1		18.6		19.2		16.4		14.7
Police-recorded crime (crimes officially recorded by the police)	5.3	5.6	5.5	5.3	5.1	5.0	4.6	4.5	4.5

*From 1998 on, totals were calculated on the fiscal year. Before 1998, calculations were based on the calendar year.

[†] The British Crime Survey was conducted only in odd-numbered years during the period 1991–99.

Sources: Home Office, "The 2000 British Crime Survey," *Home Office Statistical Bulletin 18/00*, London: Home Office, 2000; and Home Office, *Criminal Statistics England and Wales, 1999*, London: HMSO, 2000.

of the population aged 16 years and older (19,411 persons in 2000) and ascertains the number of personal and household victimizations. Police-recorded crimes include only those offenses reported to and recorded by the police.

The two data sources show similar movements in crime rates during the past 10 years. Crime increased during the first half of the 1990s, peaking in 1992 for police-recorded crime and 1995 for BCS-reported crime victimization, then declining. Although the BCS covers far fewer offense types than police-recorded crimes, the number of self-reported victimizations it records far exceeds the number of crimes recorded by the police.

The findings of the 1998 BCS on self-reported drug misuse (see exhibit 3) show that one-third (32 percent) of the general population ages 16–59 had tried an illicit drug at some time in their lives. About 11 percent had done so in the previous 12 months.[1] The most widely used prohibited drugs were cannabis (marijuana) and amphetamines. One-quarter (25 percent) of the general population had used cannabis and 10 percent had used amphetamines at some time in their lives. Use of other drugs—e.g., LSD, mushrooms, cocaine, and heroin—was much less common in the general population. No more than 5 percent of the population reported that they had ever used any of these drugs.

Exhibit 3. Self-Reported Drug Use Ever and in the Past 12 Months in England and Wales, Population Ages 16–59, 1998

Substance	Ever (%)	In the past 12 months (%)
Any drug	32	11
Amphetamines	10	3
Cannabis	25	9
Cocaine	3	1
Crack	1	<1
Ecstasy	4	1
Heroin	1	<1
LSD	5	1
Magic mushrooms	5	1
Methadone	<1	<1
Temazepam*	3	1
Solvents	2	<1

*In the British Crime Survey 1998 report, this category includes other tranquilizers.

Source: Ramsay, M., and S. Partridge. *Drug Misuse Declared in 1998: Results From the British Crime Survey*, London: Home Office, 1999.

II. Description of Criminal Justice System, Arrest Process, and Detention System

The criminal justice system in England and Wales comprises the police service, prison service, probation service, judiciary, and various criminal justice agencies, including the Crown Prosecution Service and the legal profession.[2] The criminal justice system costs more than £12 billion per year to run. Nearly two-thirds of all Government expenditures for the criminal justice system is spent on the police.

The process of arrest and detention is specified by the Police and Criminal Evidence Act of 1984 and its later additions and amendments.[3] A police officer may arrest without a warrant anyone who is guilty of an arrestable offense, who the officer has reasonable grounds to suspect is guilty of such an offense, or who is about to commit an offense.

Arrestees are held and processed in a section of a designated police station known as the custody suite or custody block. The custody suite looks like a small prison. At the entrance is a custody desk, which is often slightly elevated. This desk is staffed by two or three custody staff, usually one or two custody sergeants and a police or civilian jailer. Computer terminals on the custody desk are used to monitor arrestees and obtain Police National Computer (PNC) checks.

The rest of the custody block includes routine cells designated for use by males and females or adults and juveniles and special cells designated for use when special surveillance or special conditions are required. Two or three designated interview rooms and some nondesignated interview rooms are usually available for solicitors and other outsiders to use. The block also includes a doctor's room, a photography and fingerprinting room, an intoximeter room, a kitchen, lavatories, offices, and storage rooms.

All NEW–ADAM surveys are conducted in police stations that are designated to receive and process arrestees. Their design follows Home Office guidelines, and they operate under the constraints of the Police and Criminal Evidence Act of 1984 and its Codes of Practice.[4] This design and these constraints strongly affect the research process and how interviews are conducted.

In conducting interviews and drug testing, researchers must take note of the Act's requirements that all persons in custody must be dealt with expeditiously and released as soon as the need for detention ceases to apply. The key issue is whether interviewing and collecting voluntary urine samples from arrestees delays official processing. The primary method used to avoid delaying proceedings is to approach arrestees shortly after their arrival at the police station and before their police interrogation begins.

A second legal problem concerns disclosure. The law relating to disclosure requires the prosecution to make available to the defense any evidence on which it intends to rely in making its case. Information collected during the interview, the urine sample, and the urinalysis results could conceivably be used as evidence in court. In England and Wales, the Crown Prosecution Service has agreed not to use the materials collected as evidence. This does not, however, constrain the defense from asking for them.

III. Status of National Drug Policy

National drug policy in the United Kingdom, including England and Wales, is determined primarily by the U.K. Anti-Drugs Coordinator in collaboration with the Government. In April 1998, the Government published a parliamentary paper, *Tackling Drugs to Build a Better Britain*, based on a report by the U.K. Anti-Drugs Coordinator. This paper defines the Government's strategy on drugs for the subsequent 10 years.[5] The paper reported the four main elements of the strategy:

- To help young people resist drug misuse.
- To protect communities from drug-related antisocial and criminal behavior.
- To enable people with drug problems to overcome them.
- To minimize the availability of illegal drugs on the streets.

The U.K. Anti-Drugs Coordinator's first annual report, published in 1999,[6] announced that the Government had allocated £217 million in new money to tackle drug misuse. The report said that a large proportion of this money would be aimed at treatment and education during the next 3 years; £6 million would be allocated to research and information gathering, some of which was used to fund the NEW–ADAM Program.

Evidence exists that the Government has drawn heavily on research findings in devising its antidrug strategy and will continue to do so in evaluating the strategy's effectiveness. The *Guidance Notes* that accompanied the publication of the parliamentary paper identified the research basis for some of its initiatives.[7] Evidence of the link between drugs and crime was drawn in part from the report on the first developmental stage of the NEW–ADAM Program.[8] Program results will be used as a performance measure of at least one of the Government's key objectives—to reduce opiate and cocaine use among arrestees—and possibly other objectives, including the key objective to reduce the proportion of high-rate drug-misusing offenders.[9]

IV. Background History to I–ADAM in England and Wales

The NEW–ADAM Program can be traced to the launch of the previous Government's antidrug strategy in a parliamentary paper published in 1995.[10] The paper stressed the need to generate high-quality indicators of drug misuse and develop research on the connections between drug misuse and crime. The successor strategy, published by the Government in 1998, continued this focus on the effective monitoring of drug misuse and the need for appropriate performance measurements.[11]

In 1995, the Home Office commissioned the University of Cambridge to conduct a feasibility study on interviewing and collecting urine specimens from arrestees. The project report reviewed the literature on the U.S. Drug Use Forecasting (DUF) Program—the predecessor to the ADAM Program. The review concluded that an ADAM-type program was feasible and that much could be gained from developing such a program in England and Wales.[12]

In the same year, the Home Office received funds to commission the first stage of developmental research. The primary goals of the research were to develop a procedure for interviewing and testing arrestees based on the DUF Program in the United States and to generate information about the prevalence of illegal drug use in police force areas selected as survey sites. The first stage was conducted from January 1996 to August 1997 in five locations: Cambridge, London (Hammersmith), Manchester (Trafford), Nottingham, and Sunderland. The research report was published in 1998 and showed what was widely thought to be a high level of drug misuse among current samples of arrestees.[13]

In 1998, the Home Office commissioned the next stage of developmental research, which was conducted from July 1998 to May 1999. The research aimed to build on the early developmental work and move closer to a research design that might be used as the basis for a national research program. Surveys were conducted in London (South Norwood), Liverpool, and Nottingham. Because Nottingham had been surveyed 2 years earlier, the second survey provided an opportunity to test some of the principles involved in measuring trends in drug misuse over time.

In 1999, the Government allocated funds for monitoring drug misuse. Part of these funds was set aside to develop a national, voluntary program of drug testing and interviewing arrestees. After a competitive bidding process, the research program contract was awarded to the University of Cambridge. The NEW–ADAM Program was launched on July 1, 1999.

V. Goals and Objectives of I–ADAM in England and Wales

The goals and objectives of the NEW–ADAM Program have not yet been formally stated, but some of the program's broad aims can be identified based on knowledge of current and past thinking. One aim of present and past Government antidrug strategies has been to devise a national measure of drug misuse that can be used in conjunction with other measures of illegal drug consumption. One formal goal of the NEW–ADAM Program will likely be to generate a measure of drug misuse that can monitor changes in drug misuse over time and compare prevalence rates of drug misuse across locations as ADAM does in the United States.

Another program goal is to understand more fully the connection between drug misuse and crime. One of the main reasons for devising a measure of arrestee drug misuse is to understand patterns of stability and change in drug misuse among actual and potential offenders. Of particular interest is the extent to which offenders commit crimes to fund drug misuse and to which these drug-misusing offenders contribute to the total number of crimes committed.

A third goal of the NEW–ADAM Program is to help monitor the effectiveness of the Government's overall antidrug strategy. It is expected that one or more performance targets of the strategy will be linked to information generated by NEW–ADAM. The Government is interested in monitoring the extent to which its strategy reduces the prevalence of drug-misusing, high-rate offenders, especially those who commit crimes to fund their drug use.

A fourth goal is to help monitor the effectiveness of particular initiatives within the Government's antidrug strategy. The BCS results can be used to monitor the spread and utility of the new Arrest Referral Schemes and Drug Testing and Treatment Orders. The NEW–ADAM Program would provide broad support to evaluations designed specifically to monitor these initiatives.

VI. Getting Started

A. Obtaining Funding

The NEW–ADAM Program is funded by a central Government reserve administered by the Home Office Research, Development, and Statistics Directorate. It is expected that funding will be awarded in a series of 3-year research contracts. Funds are allocated on the basis of individual contracts, so long-term funding is not guaranteed. It is hoped that the program will demonstrate its usefulness in the first 3 years and attract Government funds in the future.

B. Contracting for Data Collection

The Home Office contracted for data collection and the entire research program through the research tender to the University of Cambridge. The university directly administers data collection. A team of researchers under contract travels to each survey site and establishes a base in the area for the duration of the research.

C. Facility Access

Access to each research site was secured through a multistage process. During the first stage, under the first 3-year contract, the Home Office wrote letters to all police force areas in England and Wales to determine interest in hosting the research. The responses to these letters and information on the number of arrestees processed through the custody suites were assessed, and a sample of forces was provisionally selected. In the second stage, the Home Office contacted the police forces selected to arrange preliminary meetings. The meetings typically included representatives from the Home Office, the University of Cambridge, and senior and junior police officers in the potential host force. Following the meetings, the representatives, usually without the most senior officers, visited each force's custody suite to determine whether it was suitable for the research.

D. Knowing the Catchment Area

The catchment area of each NEW–ADAM survey is the arrest catchment area of the custody suite, which is usually an entire police division. The police usually provide researchers with a map of the area. In most cases, the catchment area has boundaries similar to those of the local government administrative areas. This is not always so, however; in these cases, it is difficult, but not impossible, to identify the demographic characteristics of the research area.

Exhibit 4 shows the catchment areas of the first eight sites surveyed during the first year of the NEW–ADAM Program. Most were based in city center police divisions that covered geographic areas of about 5–10 square miles.

E. Drug Testing Issues

The Forensic Science Service (FSS) Urine Testing Unit at Chepstow conducted screening tests for both developmental stages of the NEW–ADAM research. During the early stages of the research, FSS provided information on the correct procedures for collecting urine specimens. The FSS's collection equipment and recommended chain-of-custody procedures have been used throughout. Important decisions had to be made about the method of drug testing, including whether to use confirmatory tests in addition to screening tests, the number and nature of the package of drugs to be tested, the type of screening test, and the cutoff levels to be used to determine a positive result. To allow flexibility in making comparisons across areas, FSS was asked to provide the raw data from the screening test to allow the results to be recalculated using different cutoff levels.

F. Instrument Issues

The questionnaire used in the surveys was based on the early DUF and ADAM questionnaires. Questions were added or subtracted, and the wording of some questions was changed slightly as appropriate. The questionnaire is in continual development, and small refinements are often made. Such changes exact a penalty in time and effort at the analysis stage. Researchers embarking on I–ADAM programs need to decide when to stop revising the questionnaire. The NEW–ADAM questionnaire currently includes a main questionnaire and two followup questionnaires, each of which is administered to alternating interviewees.

Exhibit 4. NEW–ADAM Survey Sites in the First Year of the Program, 1999–2000

City	Location	Catchment area
Sunderland	City suburb	Police division
Norwich	City center	Police division
Newport	City center	Police division
Southampton	City center	Police division
Wolverhampton	City center	Police division
Bournemouth	City center	Police division
London 1	Bethnal Green	Police division
London 2	Hammersmith	Police division

Cultural applicability

English is the main spoken language in England and Wales. Relatively few cultural minority groups do not speak English either as a first or second language. Hence, it was decided to use one questionnaire designed in English. Arrestees who did not speak English were excluded from the survey.

Standardization versus local utility

The same questionnaire is used across all locations. It has not been found necessary to change the questions to adjust for local cultural differences. The issue of cultural standardization is more relevant when making comparisons across countries. Therefore, the NEW–ADAM questionnaire includes a series of core questions that are used in the ADAM surveys in the United States.

VII. Barriers to Developing I–ADAM

The development of the NEW–ADAM Program has run fairly smoothly, but a number of potential barriers had to be tackled from the outset.

The main potential barrier to setting up an I–ADAM program was obtaining the funding and the legitimacy to carry out the research. The program requires full Government political and administrative support and central, typically Government, funding.

Another potential barrier to developing the program in England and Wales was the legal issue of disclosure in court of materials collected during the surveys. To avoid researchers being subpoenaed and having the research materials used as evidence (which would be impracticable on a regular basis), it was necessary for the Crown Prosecution Service to grant the research staff immunity from disclosure. It is hoped that this reviewable privilege will be given legal status through an act of Parliament, such as the confidentiality regulations that have been codified in the United States.[14]

Another potential barrier was the cooperation of the police. It is important that police at the most senior level agree to host the research, that middle-ranking officers agree to the procedures proposed in their police stations, and that lower ranking police in the custody suite provide full access to arrestees and to the confidential materials relating to them. One sensitive issue that had to be negotiated was the need to interview arrestees shortly after their entry into the custody suite and before their police interview. Another was the need for full access to custody records and relevant information from the Police National Computer.

VIII. National Stakeholders Involved in the Development of I–ADAM

The key players in developing the NEW–ADAM Program have been the Government, the Home Office, the police, and the University of Cambridge. The Government has provided the central funds used to administer NEW–ADAM. Its interest is in monitoring drug misuse and reducing the number of crimes committed by drug-misusing offenders. The Home Office's Research, Development, and Statistics Directorate, the Government's criminal justice research arm, administers the central funds set aside for monitoring drug misuse and evaluating Government strategy. Both the Research, Development, and Statistics Directorate and the Institute of Criminology at the University of Cambridge are interested in funding and carrying out high-quality, policy-relevant research. The police are interested in generating policing policy and supporting police research related to tackling drug misuse, which is administered through the central committee of the Association of Chief Police Officers (ACPO).

Other criminal justice organizations also have an interest in the results of the NEW–ADAM Program. The Crown Prosecution Service has granted immunity from disclosure of research materials in prosecuting arrestees taking part in the program. The Forensic Science Service provides advice and conducts all screening tests on urine specimens collected as part of the research. The local police forces that host individual surveys and their local partners have an interest in using published survey reports to inform local drug strategies.

IX. Research Methods

Selection of Police Forces

The NEW–ADAM surveys use two-stage sampling. Each of 16 locations will be surveyed once every 2 years. The first stage is selecting the police force areas to be included in the research. During the first year of the program, surveys of arrestees were conducted in eight locations; during the second year, surveys will be conducted in eight additional locations. During the third year, surveys again will be conducted in the eight locations visited during the first year. If the fourth year is funded, surveys will be conducted in the eight locations visited during the second year.

For several reasons, the choice of police forces to include in the survey could not be based on random sampling. Sampling sites must process a large number of arrestees, be willing to host the research, and together make up an even geographic spread across England and Wales and across metropolitan and other urban areas. Because the level of accuracy increases with the number of survey respondents, single-site surveys can be conducted only in custody suites where a high number of arrestees are processed. In England and Wales, the number of high-volume sites is limited. The first review of police force areas identified 16 forces (15 plus the Metropolitan Police

District in London) that had custody suites with sufficient numbers of arrestees for data analysis. Additionally, surveys can be conducted only in police force areas that agree to host the research and are fully cooperative and committed to it. The chief constable has sole responsibility for agreeing to provide researchers with access to the custody areas; the Home Office cannot overrule this decision. Finally, although nearly all the eligible high-volume police force areas were selected for the program, the final choice was made on the basis of having an even geographic spread.

Selection of Arrestees for Surveys

The second stage of sampling—the selection of arrestees in the custody block—is based on random or probability sampling. The research is conducted on a 24-hour, 7-day-a-week basis for approximately 30 days. During this period, all arrestees who pass through the custody suite are potentially eligible to be selected for an interview, although certain categories of arrestees are rejected from the outset as ineligible based on the criteria shown in exhibit 5.

All remaining arrestees are eligible for interview and should be approached by an interviewer. The method adopted is a 100-percent sample of eligible arrestees, or a census. In theory, all eligible arrestees have an equal probability of being approached for an interview. In practice, there are several practical reasons why all eligible arrestees are not approached and why all arrestees approached do not agree to be interviewed or provide a specimen. These are considered as losses in terms of the sampling procedure and reduce the research contact rates (i.e., the approach rate, the interview rate, and the specimen collection rate).

Exhibit 5. Criteria for Ineligibility for NEW–ADAM Survey Interviews

Fitness for interview
- Arrestees who are persistently unfit because of drug or alcohol intoxication
- Arrestees who are persistently unfit because of ill health or physical condition

Comprehension of interview and informed consent
- Arrestees with mental disorders
- Arrestees who require an interpreter

Potential danger to interviewer
- Arrestees who may be potentially violent
- Arrestees who are deemed ineligible at the discretion of the custody sergeant or jailer

Drug-testing requirements
- Arrestees who have been in custody for more than 48 hours
- Prison transfers and arrestees who are not at liberty before entering the custody suite

Other research selection criteria
- Children and juveniles
- Arrestees who are held only for breath-test or drunkenness offenses
- Arrestees who have been interviewed previously

X. Training Issues

The NEW–ADAM Program does not recruit new researchers for each survey (although this was done in the first few surveys) or subcontract data collection. The program uses a trained and dedicated team of researchers who travel to each site. Because each site is surveyed only once every 2 years, it was not practical to set up a local team of the type used in the United States. Because England and Wales are relatively small geographically, researchers also can travel from one location to another relatively quickly.

Researchers are typically recent graduates, usually at a master's or second-degree level, who have some experience in criminology (including criminal justice) or the disciplines that make up criminology (e.g., psychology, sociology, and other social sciences). No formal policy exists on who can be recruited as interviewers, but the job advertisements usually state as qualifications an educational background in a social science and research or professional experience working with offenders or drug users. No formal policy seeks to match interviewers with arrestees in any way.

In practice, the team does not remain stable over time. Members of the team leave to pursue their careers, and new team members are recruited. These new members undergo a similar initial training before replacing outgoing members.

This method of recruiting a dedicated team of researchers has implications for training. All new recruits are required to visit program headquarters at the University of Cambridge for an intensive 1-day training session to learn the principles of interviewing and collecting specimens from arrestees. They are also given study materials to learn on their own.

Training continues through briefing and debriefing meetings for each survey. Briefing meetings are typically held at each survey location the day before a new survey starts. Debriefing meetings are typically held shortly after a survey. General issues relating to the survey site and the methods used are evaluated. Team members regularly meet individually with the program director or manager to discuss individual training needs and personal development.

XI. National Data Analytic Results

The results of both the first and second developmental stages of the research have been published in Home Office reports.[15] The results of the first annual NEW–ADAM report are currently being assembled. It is hoped that this will be published in the second half of 2001. In addition to the program reports, individual force reports are produced, but not for general distribution.

The most recent published report is the second developmental stage report, which covers four sites surveyed in 1998–99. In total, 740 arrestees were interviewed. The proportion of males in the samples ranged from 83 to 94 percent across the four survey locations. The percentage ages 16–20 ranged from 16 to 29 percent. The percentage of nonwhite arrestees ranged from 1 to 39 percent.

The surveys showed that 69 percent of arrestees across the four sites tested positive for one or more drug types (excluding alcohol) and 36 percent tested positive for two or more drugs. The most common drug identified was cannabis (49 percent tested positive), followed by opiates (29 percent), alcohol (22 percent), cocaine (20 percent), benzodiazepines (12 percent), amphetamines (12 percent), and methadone (8 percent). Liverpool had the highest prevalence rate for positive tests for both opiates (50 percent) and cocaine (40 percent).

About one-quarter (23 percent) of all arrestees admitted injecting at least one illegal drug at some time in their lives. One-fifth (20 percent) reported injecting one or more drugs in the previous 12 months. One third (33 percent) said that they were currently dependent on one or more drugs. One-fifth (20 percent) of the total sample said that they were not receiving treatment for drug dependence but would like to receive treatment. Forty-two percent of arrestees who reported using drugs in the previous 12 months believed that their drug use and crime were connected. The connection cited most often was the need for money to buy drugs.

XII. Policy Implications and Applications of Data

Policy Implications

Evidence indicates that the published findings of the first NEW–ADAM developmental stage report have already had an effect on Government policy. The guidance notes that accompanied *Tackling Drugs to Build a Better Britain* acknowledged the first report as a source of research information.[16] Evidence on the link between drugs and crime was also drawn in part from the findings of the first developmental stage report.[17] At least one of the key performance measures used to assess the effectiveness of the Government's drug strategy will be derived from the NEW–ADAM Program.

There are also general policy implications from the findings. The results show that a high proportion of arrestees admits taking more than one type of drug and test positive for more than one type of drug. A notable proportion tests positive for opiates, including heroin, and cocaine, including crack. Almost half of drug-misusing arrestees state that their drug use and offending are connected, and many of the measures of drug use correlate strongly with many of the measures of crime.

Data Applications

NEW–ADAM data will be used to monitor drug misuse trends among arrestees to inform Government drug policy. The data will also be used to evaluate both the Government's drug strategy as a whole and individual programs.

These data have many other potential applications. The program has identified a range of problems experienced by arrestees, including health, housing, educational, and employment problems. It may be possible to monitor these problems and perhaps take advantage of the arrestees' contacts with the criminal justice system to address them.

The data also can be applied to develop fundamental knowledge about drug misuse and crime and how they are linked. Basic knowledge can be generated from ADAM surveys about drug use patterns, drug epidemics, injecting behavior, health issues, ownership and use of guns, dependency and treatment, and arrestee lifestyles. This knowledge can advance the general understanding of drug misuse and its related problems.

The data generated by the I–ADAM partnership can be used to compare international trends in drug misuse. The program allows for monitoring changes in drug misuse across countries and over time. This evidence can then be used to inform national and international drug policies.

Notes

1. Ramsay, Malcolm, and Sarah Partridge, *Drug Misuse Declared in 1998: Results From the British Crime Survey*, London: Home Office, 1999.

2. Home Office, *Digest 4: Information on the Criminal Justice System*, London: Home Office, 1999.

3. Home Office, *Police and Criminal Evidence Act 1984*, London: Her Majesty's Stationery Office [HMSO], 1986.

4. Home Office, *Police and Criminal Evidence Act 1984: Codes of Practice*, London: Home Office, 1995.

5. Her Majesty's Stationery Office, *Tackling Drugs to Build a Better Britain: The Government's 10-Year Strategy for Tackling Drug Misuse*, London: HMSO, 1998 [hereinafter *Tackling Drugs: The Government's 10-Year Strategy*].

6. U.K. Anti-Drugs Coordination Unit, Cabinet Office. *Tackling Drugs to Build a Better Britain: United Kingdom Anti-Drugs Coordinator's Annual Report 1998/99*. London: Central Office of Information, 1999. Ref. J99–4196/9905/040.

7. Her Majesty's Stationery Office, *Tackling Drugs: The Government's 10-Year Strategy for Tackling Drug Misuse* (see note 5).

8. Bennett, Trevor, *Drugs and Crime: The Results of Research on Drug Testing and Interviewing Arrestees*, London: Home Office, 1998 [hereinafter *Drugs and Crime: The Results of Research*].

9. Her Majesty's Stationery Office, *Tackling Drugs to Build a Better Britain: United Kingdom Anti-Drugs Co-ordinator's National Plan 2000/2001*, London: HMSO, 2000.

10. Her Majesty's Stationery Office, *Tackling Drugs Together: A Strategy for England 1995–98*, London: HMSO, 1995.

11. Her Majesty's Stationery Office, *Tackling Drugs: The Government's 10-Year Strategy (see note 5)*.

12. Bennett, Trevor, *A Feasibility Study of Drug Testing of Arrestees in England and Wales*, Report to the Home Office, Cambridge: University of Cambridge, Institute of Criminology, 1995.

13. Bennett, *Drugs and Crime: The Results of Research (see note 8)*.

14. 28 Code of Federal Regulations Part 22.

15. Bennett, *Drugs and Crime: The Results of Research* (see note 8); Bennett, Trevor, *Drugs and Crime: The Results of the Second Developmental Stages of the NEW–ADAM Programme*, London: Home Office, 1998.

16. Her Majesty's Stationery Office, *Tackling Drugs to Build a Better Britain: The Government's 10-Year Strategy for Tackling Drug Misuse, Guidance Notes*, London: HMSO, 1998.

17. Bennett, *Drugs and Crime: The Results of Research*.

MALAYSIA

Visweswaran Navaratnam
Vicknasingam Balasingam
Hilal Hj. Othman

Visweswaran Navaratnam is a professor at the Center for Drug Research, Universiti Sains Malaysia, where **Vicknasingam Balasingam** and **Hilal Hj. Othman** are members of the project staff.

I. Overview of Demographic Profile of Nation

Malaysia was created in 1963 through the merger of Malaya (which became independent of Britain on August 31, 1957), the former British Singapore (which later seceded from Malaysia in 1965), and the provinces of Sabah and Sarawak in northern Borneo. The country is divided into two main regions 650 kilometers apart, separated by the South China Sea: Peninsular Malaysia and East Malaysia (Sabah and Sarawak). Peninsular Malaysia borders Thailand to the north and Singapore to the south. Sabah and Sarawak border Kalimantan (the Indonesian part of Borneo) and Sarawak surrounds the tiny enclave of Brunei. The peninsula's west coast borders the Andaman Sea (Strait of Malacca). The east coast of the peninsula, Sabah, and Sarawak all adjoin the South China Sea.

Malaysia has a land area approximately the size of New Mexico and a population of more than 23 million. Its capital, Kuala Lumpur, has a population of more than 1 million. The country's three largest ethnic groups are Malay, Chinese, and Indian, and the main languages spoken there include Bahasa Malaysia (the official language), English, Chinese dialects, Tamil, and indigenous dialects. Malaysia has 13 States and 2 Federal Territories, one of which contains the capital. Exhibit 1 shows demographic information for Malaysia.

Several decades of sustained economic growth and political stability have made Malaysia one of Southeast Asia's wealthiest countries. It ranks among the medium nations (61st) in worldwide human development. Exhibit 2 shows economic and development information for Malaysia.

Penang, a relatively small but economically developed island State located in the Andaman Sea off the northwestern coast of the Malay peninsula, was chosen as the site for Malaysia's I–ADAM pilot program. The State was chosen not only for convenience but also because it had a comparatively high rate of drug use and the crimes committed there are similar to those in Malaysia's larger cities.

As of 1998, the main drug abused in Malaysia was heroin (54.7 percent of recorded cases of drug abuse). In Penang, 557 cases of heroin abuse were reported in the first 3 months of 2000. In 1998, 4,821 cases of substance abuse[1] were reported in Penang, most of which were among 20- to 39-year-olds.[2] From January to November 1999, Penang recorded 4,959 drug users, more than any other State.[3] These high prevalence rates indicate the need for appropriate intervention strategies to address substance abuse.

Little is known about the prevalence of alcohol abuse in Malaysia. Only recently was a law passed against drunk driving. Enforcement agencies in Penang rarely test arrestees' alcohol levels. ADAM could help clarify the extent of the alcohol abuse by obtaining basic background information.

II. Description of Criminal Justice System, Arrest Process, and Detention System

Malaysia is a federal parliamentary democracy with three branches: executive, judicial, and legislative. The head of state, the Yang di-Pertuan Agong (paramount ruler), is a constitutional

Exhibit 1. Basic Demographic Characteristics of Malaysia

Geographic area	330,113 km^2
Estimated population	23.26 million (as of November 22, 2000)
Ethnic distribution	Bumiputra (Malay), 58.1%; Chinese, 24.3%; Indian, 6.9; others, 3.2%; noncitizens, 7.5%
Religious affiliation	Islam (official), Buddhism, Daoism, Hinduism, Christianity, Sikhism
Gender distribution	Males, 51.2%; females, 48.8%
Age distribution	0–19 years, 43.2%; 20–39 years, 33.0%; 40–59 years, 17.7%; 60–75+ years, 6.1%
Life expectancy	70.2 years (male); 75.0 years (female); 72.2 years (average)
Population growth rate	2.3%
Birth rate	24.5 per 1,000
Total fertility rate	3.29 children per woman
Death rate	5.25 per 1,000
Literacy rate (age 15 and older)	86.4% (total); 82.0% (female); 90.7% (male)
Languages	Bahasa Malaysia (official), English, Chinese dialects (principally Cantonese and Mandarin), Tamil, Telegu, Malayalam, Panjabi, Thai, and indigenous languages
Crime rate	820 per 100,000
Drug use rate	734 per 100,000
Urbanization rate	NA
Political subdivisions	13 States and 2 Federal Territories

Sources: *National Census Data*, Kuala Lumpur: Ministry of Home Affairs, 2000; Central Intelligence Agency, *The World Factbook: Malaysia*, Washington, DC: Central Intelligence Agency, 2000; United Nations Development Programme, *Human Development Report 2000*. New York: Oxford University Press, 2000.

monarch who is elected to a 5-year term by and from among the hereditary rulers of nine States. The Parliament (Parlimen) has two houses: the lower house, the Dewan Rakyat, to which representatives are elected, in principle, to 5-year terms, and the upper house, the Dewan Nagara, to which senators are appointed by the paramount ruler and the state legislatures for 3-year terms. The leader of the party who wins a plurality of seats in the House of Representatives becomes the prime minister and head of the Government. Judges of the Supreme Court are appointed by the paramount ruler on the advice of the prime minister.[4]

In colonial times, Malaysia was confronted with the problem of opium use. The British colonial Government introduced the Dangerous Drug Ordinance and Poisons Ordinance in 1952. Malaysia's current antidrug laws, especially those dealing with drug sales, are among the world's toughest. Anyone arrested with more than 15 grams of heroin faces a mandatory death sentence. Today, Malaysia has five main statutes on substance abuse:

- Dangerous Drug Act, 1952.
- Poisons Act, 1952.
- Drug Dependence Act (Treatment and Rehabilitation), 1983.
- Dangerous Drug Act (Special Preventive Measures), 1985.
- Dangerous Drug Act (Forfeiture of Properties), 1988.

Exhibit 2. Economic Information for Malaysia

Gross domestic product (GDP)	US$171 billion
GDP growth rate	5.8%
GDP per capita	$3,238
GDP composition by sector	Industry, 46%; services, 42%; agriculture, 12%
Major export partners	United States, 23%; Singapore, 16%; Japan, 11%; Hong Kong, 5%; Netherlands, 5%; Taiwan, 5%; Thailand, 3%
Major export commodities	Electronic equipment, petroleum and liquefied natural gas, chemicals, wood and wood products, rubber, textiles
Labor force	9.3 million
Labor force composition by sector	Manufacturing, 27%; agriculture, forestry, and fisheries, 16%; local trade and tourism, 17%; services, 15%; government, 10%; construction, 9%
Poverty rate	15.5%
Unemployment rate (as of June 2000)	3.3%
Human development index (1998)*	0.771

*The human development index measures a country's average achievement in three basic dimensions of human development: a long and healthy life, knowledge, and a decent standard of living. The index is a composite index of three variables: life expectancy at birth; educational attainment (adult literacy and the combined primary, secondary, and tertiary enrollment ratio); and per capita GDP (measured as purchasing power parity in U.S. dollars). Countries are grouped into three categories: high human development (≥0.800); medium human development (0.500–0.799); and low human development (<0.500). See United Nations Development Programme, *Human Development Report 2000*. New York: Oxford University Press, 2000.

Sources: *National Census Data*. Kuala Lumpur: Ministry of Home Affairs, 2000; Central Intelligence Agency, *The World Factbook: Malaysia*. Washington, DC: Central Intelligence Agency, 2000; United Nations Development Programme, Human Development Report 2000. New York: Oxford University Press, 2000

When someone is detained for an offense, a remand order is obtained from the court within 24 hours after the arrest to facilitate investigations. An arrestee can be detained under remand for up to 14 days, after which formal charges must be made or the arrestee released.

Most arrestees are tested for drug use. If an arrestee tests positive for drugs, a medical officer will determine whether the arrestee is drug dependent, usually through urine tests and physical examinations. If an arrestee is found to be drug dependent, the National Narcotics Agency (NNA) recommends the most appropriate treatment to the court. The two types of treatment are community supervision and institutional rehabilitation, both of which normally last for 2 years, and both of which are discussed in more detail in Section III. First-time drug offenders and those who are found to have strong family support are normally recommended for community supervision. Habitual and hardcore drug users are sent for institutional rehabilitation. After 2 years of institutional rehabilitation, the drug user is placed under community supervision for another 2 years. If the supervision rules are broken during this period, the offender can be sent to prison for not less than 3 years.

III. Status of National Drug Policy

In 1996, the NNA approved the new strategies of the National Drug Policy, in the following priority order:

- Prevention.
- Enforcement.
- Treatment and rehabilitation.
- Regional and international cooperation.

Prevention, enforcement, and treatment and rehabilitation are the three main strategies; international cooperation is sought to support these strategies.

Prevention

Prevention is a long-term strategy aimed at protecting individuals from drugs. Prevention programs emphasize individual resistance to drug use and efforts to create an environment that will protect individuals and community from drug abuse. These efforts include—

- Encouraging young people to reject the drug lifestyle.
- Changing society's perception of and attitude toward drugs.
- Mobilizing the community in drug abuse prevention programs.
- Providing alternatives to drugs and encouraging healthier lifestyles.

This strategy has two components: primary and general prevention. Primary prevention involves developing an individual's inner strengths through education in positive religious, moral, cultural, and social attitudes and values. Programs that have been implemented include preventive drug education in schools, teacher training, and guidance and counseling services. General prevention involves the community in drug control and prevention. General prevention strategies have been implemented through information campaigns, publicity programs, and workplace prevention initiatives.

Enforcement

Enforcement has three main components: interdiction, legislation, and intelligence. Interdiction aims to reduce the supply of drugs on the market by preventing the smuggling of illicit drugs into the country. The strategy emphasizes strengthening and intensifying enforcement measures to curb the entry of illicit drugs by land, sea, and air. At the same time, aggressive efforts are carried out continuously to stop trafficking and sales of illicit drugs within the country.

Legislation complements prevention efforts by imposing severe penalties that make it dangerous to possess, supply, traffic in, or abuse drugs. As noted above, Malaysia's drug laws are among the world's toughest.

Intelligence from inside and outside the country can be a major asset in combating syndicates and individuals involved in drug smuggling. Intelligence gathering has been strengthened by the establishment of a Special Intelligence Team in border areas with Thailand and in the States of Sabah and Sarawak, which can carry out joint operations with the intelligence services of other countries to curb the supply of illicit drugs at the source.

Treatment and Rehabilitation

This strategy has two goals: eliminate drug dependency and prevent recidivism. To eliminate drug dependency, addicts are ordered to participate in one of two treatment and rehabilitation programs: community supervision or institutional rehabilitation. Community supervision is rehabilitation in a community setting. Drug users under community supervision must report weekly to the district NNA officer for urine testing and counseling. Anyone found to be using drugs will be sent into institutional rehabilitation.

Institutional rehabilitation occurs at one of 25 residential drug rehabilitation centers. These centers treat drug addiction through a combination of "cold turkey" methods, counseling, and vocational therapy. Treatment normally lasts 2 years.

Antirecidivism efforts focus on strengthening the attitudes and behavior of ex-addicts to prevent relapse and to enable them to live drug-free lives. The programs implemented include supervision, counseling, and social support programs at district-level service centers operated by the NNA.

International Cooperation

The national drug strategy needs international support and cooperation. Its objectives are to enhance regional and international cooperation in efforts to control and prevent drug abuse and trafficking and to strengthen national efforts toward international drug control and prevention.

IV. Background History to I–ADAM

The relationship between alcohol and other drugs (AOD) and crime has not been studied much in Malaysia. Although drug issues manifest themselves at every level of the criminal justice system, from street crime to international trade in drugs, the relationship between AOD and crime is difficult to specify because most crimes result from a combination of situational, personal, cultural, and economic factors. There are not enough data currently available to provide definitive evidence to law enforcement authorities, service providers, and researchers of the link between drug use and the drug trade and crime in general.

The National Institute of Justice, the research arm of the U.S. Department of Justice, invited representatives of the Center for Drug Research, Universiti Sains Malaysia, to participate in the first International Arrestee Drug Abuse Monitoring (I–ADAM) conference in Chicago in 1999. As a result of that meeting, Malaysia decided to implement an ADAM pilot project in Penang. The project designers hope to show the feasibility of implementing the ADAM mechanism as an ongoing activity in Malaysia. By measuring AOD use among arrestees, they hope to take the first steps to reduce AOD-related crimes, which is an important component to reducing crime in Penang.

V. Goals and Objectives

The general aims of the ADAM project are as follows:

- To assess the feasibility of implementing and sustaining an ADAM project in the State of Penang.
- To investigate the relationship between alcohol and other drugs and crime.
- To better understand the prevalence of drug use among hidden populations (subsets of the general population whose membership cannot be readily distinguished or enumerated based on existing knowledge or sampling capabilities) and its relationship to crime.[5]
- To disseminate the data collected from the project to local health and law enforcement agencies.
- To participate in an international network of ADAM sites in order to share information and develop new methods of measuring arrestee drug use and of applying the information learned to the problems of reducing drugs and crime.

The project's specific objectives are to—

- Identify the extent of AOD use among recent arrestees in Penang.
- Identify patterns of substances abuse among arrestees in Penang.
- Investigate the extent of alcohol and substance abuse among cases referred to the hospital by police (e.g., automobile accidents where alcohol or drugs may have contributed).
- Assess the influence of environmental conditions on AOD use.

- Obtain a profile on arrestees (e.g., sociodemographic data, arrest history, types of offenses).
- Investigate the correlation between AOD use and crime.
- Understand arrestees' perceptions of crime and attitudes toward law enforcement.
- Test U.S. survey instruments in local populations.
- Develop and validate reporting techniques such as self-report and biological measures (urine testing).
- Develop specific intervention techniques for various sectors (e.g., health, criminal justice, welfare, corrections).

VI. Getting Started

A. Obtaining Funding

Funds for the pilot phase were obtained through a grant application by the NNA to the national research program known as IRPA (Intensified Research in Priority Areas). IRPA supports research that it believes is a national priority and has the potential to contribute to the formation of local policies.

B. Data Collection

NNA officers and a research officer have been appointed to collect data from participating police stations. Police officers will notify the research team daily on how many arrests were made the day before, and the data collectors will visit the police station daily to collect data from arrestees.

C. Facility Access

To ensure that the project would be implemented in accordance with ADAM procedure, the Center for Drug Research, Universiti Sains Malaysia, collaborated with NNA and the police in Penang. Once this collaboration was established, access to facilities was arranged by NNA officers and the local police.

The plan is that the police stations involved in the study will call the NNA office daily to tell how many arrestees were brought into each station the day before. The research officer from the Center for Drug Research will randomly pick every 10th arrestee brought in the day before to be interviewed. The research officer will conduct the interview at the police station within 48 hours of the arrest. A voluntary urine test will also be carried out to determine whether the interviewee recently used drugs.

D. Knowing the Catchment Area

Penang has two police divisions, the northeast and southwest divisions. The northeast division was selected as the catchment area for this study. The division has 4 teams with a total of 12 police stations, all of which will participate in the study. Exhibit 3 lists the participating police stations.

E. Drug Testing Issues

Urinalysis

The validity of interviews and self-reported data can be verified by chemical drug testing of biological specimens (urine). Litmus paper is being used for the initial test on the urine sample. The litmus paper used is in accordance with U.S. National Institute on Drug Abuse (NIDA) standards; its detection sensitivity is 50 mg/mL or greater. Results are received in 5 minutes. If the litmus paper shows a positive result, the urine is sent to a Government pathologist for confirmatory tests.

Because the police collect the urine specimens, arrestees may not give the specimen voluntarily, which introduces bias into the testing. The test results of arrestees who refuse to participate in the study are not examined and, hence, cannot be compared with the results of those who agree to participate.

F. Instrument Issues

Questionnaire

The ADAM questionnaire to be used in this study was translated into Bahasa Malaysia, then translated back into English as a check on the accuracy of the translation. A pretest was carried out among a few arrestees to determine the clarity of the translation.

Exhibit 3. List of Police Stations Participating in Penang ADAM Project

Team A	Team B	Team C	Team D
Palau Tikus	Petani Road	Air Itam	Sungai Nibong
Georgetown	Dato Keramat	Bandar Baru	Jelutong
Central		Kampang Baru	
Beach Street		Tanjong Tokong	

VII. Barriers to Developing I–ADAM

The research team established a collaborative effort to ensure that administrative barriers do not hamper the implementation of the project. Because this study is a pilot project, the actual barriers can be identified only after the study is completed. Several potential problems can be noted, however, including the following:

- As noted above, the police will administer the urine testing, which may reduce participation by interviewees.
- A sample of every 10th arrestee may yield too few participants to get a truly representative sample of arrestees (see section IX for a discussion of the potential sample size).
- Because the arrestees will have already been in jail for a day when the researchers are notified of the arrest, the researchers have less opportunity to interview the arrestees before 48 hours have passed since the arrest. This could become more of a problem if several stations process a larger than usual number of arrestees at once or if not enough research officers are available to handle all 12 stations at a given time.

VIII. National Stakeholders Involved in the Development of I–ADAM

The Center for Drug Research, Universiti Sains Malaysia, is collaborating with the NNA to implement the ADAM project in Penang. Researchers from the center and NNA officers also have met with the police to brief them and to obtain their permission to conduct the study. As the source of the funding for the study, IRPA is a stakeholder, as are the national and local entities that form IRPA's constituencies.

IX. Methodology

Study Design

The Penang I–ADAM project is a 1-year pilot project with a cross-sectional design to assess the feasibility of developing an ADAM system.

Sample Size

As the study is still ongoing, the actual sample size is not yet known. It is hoped that approximately 150 arrestees will be interviewed and tested in the first 6 months of the study. The arrestees are being selected randomly for interviews and testing (see section VI, D).

Men 18 years of age and older are being selected for the study. Persons younger than 18 are defined as juveniles in Malaysia, so they come under a different jurisdiction. Because only a small fraction of those arrested in Penang for AOD offenses are women, it would not be practical to include women in the sample.

All arrestees in the selected police stations who have been in custody less than 48 hours and are not considered violent or mentally ill are eligible to be recruited for the study (with their consent). The 48-hour limit will ensure the validity of the urinalysis. Exhibit 4 lists the crimes that are to be included in the sample.

The following arrestees will not be eligible for inclusion in the study:

- Children or juveniles.
- Persons who are mentally ill.
- Persons who are considered violent.
- Persons who have been detained longer than 48 hours.

X. Training Issues

The research officer from the Center for Drug Research and the officers from NNA were trained by the principal investigators on conducting interviews and the survey methodology. In addition, police officers were briefed about the project to ensure that they understand its objectives. The research officers from the Center for Drug Research and NNA have experience dealing with the addict population and are familiar with the ADAM interview techniques.

Exhibit 4. List of Offenses to Be Sampled in ADAM Project

Abduction	Indecent assault
Arson	Intercourse with an underage girl
Assault	Kidnaping
Carjacking/truck hijacking	Malicious damage to property
Common burglary	Motor vehicle theft
Driving under the influence of alcohol or other drugs	Murder or attempted murder
Embezzlement	Public violence
Fraud and misappropriation of funds	Rape
Illegal possession of firearms	Robbery with aggravated circumstances
Ill treatment of children	

XI. National Data

No national data are available yet.

XII. Policy Implications and Application of Data

The I–ADAM pilot project will complement the existing Asian Multicity Epidemiology Network and the National Drug Information System. The I–ADAM project is unique because it will provide data on drug use among the arrestee population; the two existing programs provide data from the addict population. The data from this study will provide an indicator to policymakers of the extent of drug and alcohol use among those arrested for crimes. It is hoped that this study will be used and accepted as an indicator and, in the long run, will influence existing policy regarding drugs and crime.

Notes

1. In Malaysia, substance abuse (known as "daddah" in Bahasa Malaysia) does not refer to alcohol abuse or dependency.

2. National Narcotics Agency, *National Narcotics Report,* Kuala Lumpur: Ministry of Home Affairs, 1998.

3. National Narcotics Agency, *National Drug Information System,* Kuala Lumpur: Ministry of Home Affairs, 2000.

4. Central Intelligence Agency, *The World Factbook: Malaysia,* Washington, DC: Central Intelligence Agency, 2000.

5. See Weibel, W.W., "Identifying and Gaining Access to Hidden Populations," in *The Collection and Interpretation of Data From Hidden Populations,* ed. E.Y. Lambert. NIDA Research Monograph 98. DHHS Publication No. (ADM) 90–1678. Rockville, MD: U.S. Department of Health and Human Services, National Institute on Drug Abuse, 1990.

THE NETHERLANDS

Peter Klerks
Janine Plaisier
Willemijn Garnier

At the time this chapter was written, **Peter Klerks** was a researcher for Eysink, Smeets, and Etman in The Hague, the Netherlands. He currently works for the Dutch Police Academy. **Janine Plaisier** and **Willemijn Garnier** were research coordinators for WODC, the research and documentation center of the Ministry of Justice, The Hague, the Netherlands. **Willemijn Garnier** now works for the Ministry of Justice in The Hague.

I. Overview of Demographic Profile of Nation

The Netherlands, a small country in Western Europe bounded by Germany to the east, Belgium to the south, and the North Sea to the west, is one of the most urbanized and densely populated countries in the world. With an area of only 41,526 square kilometers (approximately twice the size of New Jersey), it had a population of about 15.8 million in 1999 (7.8 million men and 8.0 million women). There are 662,000 foreign nationals living in the country, and 17.1 percent of the Dutch population are first- or second-generation immigrants. Basic demographic information is provided in exhibit 1, and a more detailed demographic breakdown is given in exhibit 3.

A prosperous country, in 1999, the Netherlands had a GDP of 842 billion guilders ($365.1 billion), an unemployment rate of 3.3 percent, a poverty rate (the percentage of families with an income at or below the social minimum) of 10.5%, and a literacy rate—excluding functional illiterates (people who can read but do not understand what they read)—of 87 percent.[1] Basic economic information for the Netherlands is given in exhibit 2.

The I–ADAM pilot program in the Netherlands took place in one of the country's 25 police regions, The Hague (Haaglanden), which is the seat of the national government and has about 450,000 inhabitants.

Crime Statistics

In 1999, there were 86,587 violent crimes in the Netherlands, a 13-percent increase from the previous year and a continuation of the upward trend since 1995, when there were 64,606 violent crimes

Exhibit 1. The Netherlands: Demographic Overview

Geographic area	41,526 km^2
Population (1999)	15.7 million
Ethnic distribution (1999)	Dutch, 91%; Moroccans, Turks, other, 9%
Religious affiliation (1991)	Roman Catholic, 34%; Protestant, 25%; Muslim, 3%; other 2%; unaffiliated, 36%
Gender distribution	Males, 49%; females, 51%
Life expectancy	Males, 75.4 years; females, 81.28 years; average, 78.28 years
Population growth rate (1000)	0.6%
Birth rate	12 per 1,000 population
Total fertility rate	1.6 children per woman
Death rate	9 per 1,000 population
Literacy rate	99%
Functional literacy rate	87%
Political subdivisions	12 provinces

Sources: U.S. Census Bureau, International Database, 2000; *CIA World Factbook*, 2000

Exhibit 2. The Netherlands: Economic Information

Gross domestic product (GDP) (1999)	$365.1 billion
GDP growth rate (1999)	3.4%
GDP per capita (1999)	$23,100
GDP composition by sector	Agriculture, 3.5%; industry, 26.8%; services, 69.7%
Major export partners (1998)	EU (Germany, 27%; Belgium-Luxembourg, 13%; France, 11%; UK, 10%; Italy 6%); Central and Eastern Europe, U.S.
Major export commodities	Machinery and transport equipment, chemicals and fuels, foodstuffs, clothing
Labor force (1998)	7 million
Labor force composition by sector (1998)	Services, 73%; industry, 23%; agriculture, 4%
Poverty rate	10.5%
Unemployment rate	3.3%

Sources: U.S. Census Bureau, International Database, 2000; *CIA World Factbook*, 2000

(see exhibit 4).[2] There were 872,362 property crimes in 1999, a 4-percent increase from the previous year, but still below the 1995 figure of 903,871. The number of Drug Act crimes has increased from 3,473 in 1995 to 7,613 in 1999. This category reflects only violations of the Opium Act, according to which drugs are categorized as "hard" and "soft." Hard drugs include heroin, cocaine, and synthetic drugs such as MDMA (also known as Ecstasy or XTC); soft drugs include cannabis derivatives such as marijuana and hashish.

Drug Use Statistics in the General Population

The first National Drug Use Survey was carried out in 1997 among 22,000 persons 12 years and older using a computer-assisted personal interview. The design allows prevalence data to be broken

Exhibit 3. The Netherlands—Age Distribution of Population, 1999

Age (years)	Number of people
0–4	967,000
5–9	995,000
10–14	966,000
15–19	921,000
20–29	2,088,000
30–39	2,646,000
40–49	2,371,000
50–59	2,036,000
60–69	1,381,000
70+	1,520,000

Source: U.S. Census International Database, 2000

Exhibit 4. Official Reports of Crime to the Police

	1995	1996	1997	1998	1999
Violent crime	64,606	66,688	74,430	76,610	86,587
Property crimes	903,871	833,669	846,033	842,324	872,362
Malicious damage/public order crimes	152,480	170,412	179,780	178,430	188,242
Road traffic act	89,229	105,159	99,700	105,449	117,087
Drugs act	3,473	6,593	10,300	7,666	7,613
Firearms act	1,865	2,198	1,986	3,350	3,432
Other offenses	7,314	5,370	6,647	8,397	9,005
Total	1,222,838	1,190,089	1,218,876	1,222,226	1,284,328

Source: Department for Statistical Information and Policy Analysis, *Registered Crime in the Netherlands, 1995–1999*, The Hague: Ministry of Justice, 2000

down into four metropolitan areas (Amsterdam, Rotterdam, Utrecht, and The Hague) and five areas with different population densities. Data from the second survey, which was carried out in 2000, will be available by the end of 2001 (see exhibit 5).[3]

It is evident from exhibit 5 that cannabis was by far the most popular illicit drug both in terms of lifetime use and in terms of usage in the last month. Looking at lifetime prevalence rates, cocaine was the most popular drug, followed closely by amphetamines, Ecstasy, and hallucinogens; only 0.5 percent of the respondents had reported using any "hard drug" in the month preceding the interview. Mushrooms were used at a similarly low rate. Heroin use was so low as to be hardly measurable, but population surveys are not suitable to quantify levels of heroin use. For all drugs, the difference between lifetime and recent use indicates that most people who experiment with a drug do not continue to use it. This has been expressed in a so-called continuation rate: Some 16 percent of the lifetime cannabis users continued to use it in the past month. In contrast, continuation rates for alcohol and tobacco were 80 and 51 percent, respectively.

Exhibit 5. 1997 Dutch National Drug Use Survey (%)

Lifetime use	Amsterdam	Rotterdam	The Hague	Utrecht	Nonurban*	National average
Cannabis	36.7	18.5	20.1	27.3	10.5	15.6
Cocaine	9.5	3.4	3.4	3.6	1.0	2.1
Ecstasy	7.0	2.2	2.6	3.2	1.2	1.9
Amphetamines	6.0	2.7	2.2	2.6	1.1	1.9
Hallucinogens[†]	6.3	1.8	2.8	3.0	1.1	1.8
Mushrooms	6.6	2.4	2.5	n.a.	1.0	1.6
Heroin	1.8	0.4	0.5	0.3	0.3	0.3
Use in the past month						
Cannabis	8.1	3.3	4.2	4.2	1.5	2.5
Cocaine	1.0	0.4	0.6	0.4	0.1	0.2
Ecstasy	1.1	0.1	0.2	0.7	0.1	0.3
Amphetamines	0.3	0.1	0.2	0.3	0.1	0.1
Hallucinogens[†]	0.0	0.0	0.1	0.4	0.0	0.0
Mushrooms	0.6	0.1	0.1	n.a.	0.1	0.1
Heroin	0.3	0.0	0.1	-	-	-

Note: Percentage of the Dutch population aged 12 years and older reporting drug use in 1997.

*Defined by Statistics Netherlands as municipalities with fewer than 500 addresses per square kilometer.

[†]Including LSD, mescaline, psilocybin, 2CB, and ayahuasca and *excluding* mushrooms.

Source: National Drug Use Survey, CEDRO (Abraham et al., 1999)

If prevalence rates of last-month use are extrapolated to the general population age 12 years and older, one may obtain a "lower bound" or minimum estimate of the number of drug users in the country (see exhibit 6). This is a minimum estimate because some groups of drug users, such as homeless youth, prisoners, opiate addicts, and frequent visitors of coffee shops,[4] are likely to be underrepresented in the survey. For example, cocaine is used by the large majority of opiate users, who are generally not captured sufficiently in the sample of a population survey. The actual number of cocaine users is probably much higher.

Political System

Since 1848, the Netherlands has been a constitutional monarchy in which the government—the Council of Ministers under the leadership of the prime minister—is responsible for government policy and responsible to parliament. The Queen enjoys a position of immunity.

Parliament consists of an upper house (the First Chamber), and a lower house (the Second Chamber). The 150 members of the Second Chamber are chosen directly by the people in elections that are, in principle, held every 4 years. The Second Chamber's most important function is to supervise the government's actions. To do so, it has several powers, one of the most important of which is the right to amend bills proposed by the cabinet. The minister responsible for a proposed bill can adopt such an amendment, submit it to a vote in the Second Chamber, or reject it. If a parliamentary amendment

Exhibit 6. Minimum Estimates of the Number of Drug Users in the Netherlands, 1997

Drug	Absolute number of users*
Cannabis	323,000
Cocaine	26,500
Ecstasy	40,000
Amphetamines	13,000

*Extrapolated from last-month prevalence rates to the Dutch population aged 12 years and older.

is rejected, the lower house may bring a motion of no confidence against the minister or the cabinet, which, if carried, will force their resignation. The lower house also has the right of initiative (i.e., the right to propose bills), the right of interpellation (i.e., the right to demand clarification from a minister), the right of inquiry, and the right to accept or reject the Government's budget.

The 75 members of the First Chamber are elected indirectly for 4-year terms by the members of the country's 12 provincial councils. The First Chamber shares in the legislative function and monitors government policy. All bills that have been passed in the Second Chamber must also be approved by the First Chamber, but the First Chamber cannot propose or amend bills. Members of the First Chamber have the same rights as those of the Second Chamber.

The Netherlands has 13 government ministries, each of which is headed by a minister who is politically responsible for that ministry's policy. Each minister is supported in this task by one or occasionally two politically appointed state secretaries. Civil servants, who maintain an apolitical stance and do not lose their jobs with a change in government, assist ministers and state secretaries in their work.

II. Description of Criminal Justice System, Arrest Process, and Detention System[5]

The cornerstone of Dutch criminal law is Article 1 of the Criminal Code, which states that no offense is punishable if it does not violate a previously existing statutory penal provision. Only conduct that was punishable when the act occurred is subject to sanctions. A punishable offense is an action that meets the statutory description of the offense, is wrongful, and is due to an imputable fault. Crimes are divided into offenses (more serious crimes) and violations.

Criminal proceedings are held in phases, the first of which is prosecution in the first instance. This phase commences with the police investigation, the results of which are set out in a police report stating the nature of the offense, when it took place, and the suspect's name.

A suspect can be brought to the police station for interrogation, which can take place during a maximum of 6 hours. If the investigation requires it, the suspect can be taken into custody (only for punishable offenses for which pretrial detention is allowed) for a maximum of 3 days.[6]

Sometimes (e.g., in the case of minor violations), the police do not draw up an official report. If the police do not offer a transaction (a deal under which the offender voluntarily pays a sum of money to the Treasury to avoid further criminal prosecution and public trial), however, and decide to press charges, the official report is passed on to the Public Prosecutions Department. Once the criminal investigation has been closed, the Public Prosecutor may drop the case, settle it by means of a transaction, or serve the suspect with a writ of summons if he believes sufficient evidence exists to reach a conviction.

The trial in court concludes the first phase. In the Netherlands, there is no trial by jury; independent, professional judges who serve for life decide both the law and the facts of the case. Lay people are not involved in either the course of criminal proceedings or the appointment of judges. Hearings take place in public whenever possible. In principle, juvenile court cases are heard in closed chambers. If a punishable offense has been proved and the offender is found guilty, the judge decides which penalty or measure should be imposed and imposes a sentence.

The second phase involves appellate proceedings in either the Court of Appeal (appeal) or the Supreme Court (appeal in cassation). Both prosecution and defense have the right to bring a one-time appeal to a higher court. Appeals against a judgment given by a district court can be lodged with the Court of Appeal. Taking a case (appeal in cassation) to the Supreme Court differs from lodging an appeal because it does not result in an (entirely) new hearing of the case but serves merely to establish whether the sentence challenged is in accordance with the law and in compliance with the requirements.

The third and last phase involves the enforcement of a sentence. When appeals are exhausted, the sentence becomes final and conclusive and can be enforced. The Public Prosecutions Department is responsible for implementing the criminal sentence. It does not impose the sanctions itself, but delegates imposition to judicial or private institutions.

The sanctions contained in the Dutch Criminal Code are classified as penalties and measures. Penalties are retaliatory or deterrent in nature, while measures focus on protecting the interests of society or the perpetrator. Unlike penalties, measures need not establish a direct relationship between the gravity of the offenses committed and the severity of the legal response to it. Legal articles set the maximum sanction the judge can impose for any given offense. The most common sanctions for adults are as follows:

- Imprisonment, the most substantial and severe penalty in the Dutch Criminal Code, can be imposed only for criminal offenses. Prison sentences can be for life or of temporary duration. For any given offense, judges have the freedom to impose prison sentences within minimum and maximum limits. The maximum sentence for theft, for example, is 4 years, whereas the maximum sentence for murder is 20 years.
- Detention can be imposed for a minimum of 1 day and a maximum of 1 year and 4 months and is generally imposed for summary offenses. Detention is lighter than a prison sentence and, as a rule, is spent in a remand prison.

- A community service order (taakstraf) can also be imposed on minors who have committed a punishable offense. The accent is on first offenders and those who have committed less serious offenses. Under a community service order, an offender can be sentenced to up to 240 hours of community service (werkstraf), to follow a social training course (leerstraf), or a combination of the two. Both the judge and the Public Prosecutions Department can impose the penalty. Strictly speaking, only the community sentence imposed by a judge has the status of a court sanction.
- A fine can be imposed for all punishable offenses, even those for which the judge could impose a life sentence.

In certain exceptional cases, such as if the offender is mentally ill, the judge can impose a measure (e.g., a psychiatric hospital order or an entrustment order) instead of a penalty.

III. Status of National Drug Policy

Health issues are of primary importance in Dutch drug policy.[7] A broad approach is taken, in which attention is given to both prevention and care (demand) and combating the sale of drugs (supply). The demand for drugs is discouraged and the risks of drug use to individuals and their environment are diminished by a professional care and prevention policy. The Government seeks to prevent young people, in particular, from unthinkingly using drugs in the first place and to make medical and social assistance available to problem drug users to alleviate their plight (harm reduction). Harm reduction is most obvious in the criminal justice system in the imposition of relatively low sentences for possession of small amounts of drugs, but harsh sentences for persons involved in the drug trade. The supply of drugs is fought against by combating organized drug crime. Dutch drug policy is also aimed at maintaining public order and reducing the nuisances caused by drug use.

Assistance is given not only to "normal" drug users but to criminal drug users as well. The judicial authorities have followed the principle that drug addicts who commit minor crimes should be offered the opportunity to undergo medical treatment rather than serve a custodial sentence. This makes it easier to persuade drug addicts to participate voluntarily in special programs that help them regulate or stop their drug use and reenter society. The aim of care services is to reach as many addicts as possible to help them rehabilitate or limit the risks caused by their drug habit. Social rehabilitation is an essential element.

In investigations and prosecutions, the expediency principle is applied (i.e., a public prosecutor may decide not to prosecute if prosecution is not in the public interest). A distinction is made between hard drugs, which pose an unacceptable hazard to health, and soft drugs, which constitute a far less serious hazard. The highest priority is given to the investigation and prosecution of international drug trafficking; the possession of small quantities of drugs for personal use is accorded a much lower priority. Possession of either soft or hard drugs is considered an offense. Possession of more than 0.5 grams of hard drugs is an indictable offense. Firm action is taken in such cases. Possession of smaller amounts of soft drugs is a summary offense and therefore has low priority in policies on investigation and prosecution. The expediency principle is applied to the sale of cannabis in coffee shops to separate the markets for hard and soft drugs and to keep young people who experiment with cannabis away from hard drugs.

Combating drug-related crime is aimed at the fight against the trade and production of drugs, especially hard drugs. One action recently taken was the foundation of the Synthetic Drugs Unit, which is specifically aimed at the fight against the production of, and trade in, Ecstasy, amphetamines, and other synthetic drugs.

Research on drug use is of great importance in the Netherlands. More and more information is being gathered because of the accessibility of drug treatment services and better data registration systems. First of all, research is done to obtain information about trends in drug use, drug users, and the supply of drugs and related dangers. Studies are regularly conducted on drug use in general and among students in particular (all forms of research are applied: field research, investigative research, evaluation research and policy-oriented research). In the past few years, multidisciplinary research has been conducted into the use, effects, and risks of Ecstasy and other synthetic drugs on the basis of, for example, data obtained from tablet tests at rave clubs.

Much information is available because of field research.[8] Less information is available, however, on persons who use more than one kind of drug (including alcohol) and the relationship of such use to criminal behavior. One way to obtain more information about this could be through the I–ADAM project. The following sections describe the pilot program conducted in the Netherlands in 1999.

IV. Background History of I–ADAM in the Netherlands

In 1998, WODC, the Research and Documentation Center of the Dutch Ministry of Justice, was contacted by the U.S. National Institute of Justice (NIJ) about the ADAM project. NIJ wanted to expand ADAM internationally (in a project called International ADAM, or I–ADAM) and asked whether WODC would be interested.

A senior advisor of the Dutch Ministry of Justice and a police commissioner of the Hague police were invited to attend the first I–ADAM conference in Miami in April 1998. On returning to The Hague, they decided to join forces and managed both to secure both the funds to finance a feasibility study (from WODC) and to persuade the Hague regional police force to serve as a testing ground for an ADAM experiment. WODC could then start a feasibility study and a pilot.

V. Goals and Objectives of I–ADAM in the Netherlands

As previously stated, the Netherlands has a well-established tradition of research on drug consumption patterns. The initiators of the new I–ADAM pilot program would have to prove that it could bring new insights to the field. Starting carefully, they decided that a feasibility study would first have to show that the major stakeholders were prepared to allow a "live" experiment involving arrestees. The burgomaster of The Hague, who is the head of the regional police force; senior decisionmakers from the Public Prosecutions Department and the Ministry of Justice; the chief commissioner of the police; health care officials; and the Bar Association (orde van advocaten) were all consulted. With the exception of the Bar Association, they reacted positively to the idea. The Bar Association demanded a very strict privacy regime under which each arrestee would have to confirm his willingness to participate in the program by signing an informed consent form.

Another objective of the feasibility study was to ascertain what kind of information the stakeholders desired and expected to learn from the I–ADAM study. It turned out that most respondents thought that the innovative approach could give new insight into the drug consumption patterns of people who normally remain outside the established drug research studies, either because they do not feel they have a drug problem or because they have become so marginalized that even field workers cannot reach them anymore. The prospect of discovering a connection between specific forms of criminal behavior and drug consumption habits also stimulated enthusiasm, as did the possibility that new trends could be identified at an early stage and that of making international comparisons through the I–ADAM research network. The realistic and most frequently expressed opinion, however, was that the experiment was aimed primarily at determining—

- Whether such a method could be carried out with success.
- Whether I–ADAM in the Netherlands would produce policy-relevant information.
- Which barriers would be encountered and how could they be overcome.

The actual results of the interviews and urine tests would be less important than the process evaluation of the experiment itself and the experiences of the interviewers involved. The experiment would have to produce insight on how I–ADAM could best be implemented on a large scale in this and other facilities throughout the country.

The feasibility study (and a mini-pilot that tested 11 arrestees) took place in December 1998. The pilot experiment was scheduled to start in the spring of 1999, but after the Bar Association objected, it was necessary to review certain procedural arrangements to comply with its demands. In the end, all the Bar Association's objections were satisfied and the experiment was allowed to proceed under strict guarantees of anonymity and voluntary participation. Preparations were completed in the summer, and on August 16, 1999, the first arrestees were interviewed.

The intention was for the pilot study to obtain 250 completed interviews in an estimated 6 weeks of interviewing. After a full month of interviewing, however, fewer than 100 interviews had been completed, mainly because there was limited time during the day when arrestees could be interviewed (only during the morning, after showering). The supervisory commission therefore decided to stop after the 125th interview. On September 27, after 125 interviews were held, the gathering of data was terminated, and evaluation and data analysis began.

VI. Getting Started

A. Obtaining Funding

Funding for the I–ADAM pilot came from the research budget of WODC, the Dutch Ministry of Justice's in-house research department. WODC's research in the broad field of criminal justice is carried out in part by its own researchers and in part by contracted researchers from universities and research companies. WODC's director decided to fund the pilot because the objectives of a Dutch I–ADAM project fit into WODC's research program and the project could produce interesting results for Dutch policymakers on drug-related subjects.

B. Contracting for Data Collection

The work for the feasibility study and mini-pilot to establish whether the I–ADAM concept would be acceptable to stakeholders in the Netherlands was granted to Eysink, Smeets, and Etman (ES&E), an independent research and consultancy firm specializing in crime prevention and security matters that has an extensive track record researching difficult and inaccessible groups.

C. Facility Access

Since the Hague regional police force had been involved in the project from the beginning, gaining access to the cell block facilities posed no problem at all. The chief of the cell block staff was cooperative and interested in the experiment, as were most of his staff. In the interest of the arrestees' privacy and the professional detachment required of the research staff, the initially very informal atmosphere had to be restrained by a set of guidelines involving strict instructions not to discuss anything related to the interview subjects. Since the normal daily routine in the cellblock is based on an efficient and optimal use of personnel and requires a tight schedule, however, it was not always easy to get access to the arrestees themselves.

D. Knowing the Catchment Area

Arrestees were brought in from the entire region covered by The Hague regional police force. Since not all the arrestees in the region participated in the I–ADAM pilot program, the results cannot be compared with the general population of arrestees in this area. The following statistics give some general indications of the area's total arrestee population. In 1998, 14,605 persons, 15 percent of whom were female, were arrested for one or more offenses. Approximately 60 percent of arrestees in The Hague were of Dutch nationality, whereas 10 percent were from Suriname, 4 percent were Turkish, 4 percent were from the Netherlands Antilles, and 7 percent were Moroccan. The mean age was 33 years. The main offenses were theft and burglary without violence (53 percent), traffic offenses (38 percent), criminal damage (32 percent), and violent crime (27 percent).[9] In 1998, 1,007 hard-drug addicts (7 percent of all arrestees) were arrested for one or more offenses, accounting for 15 percent of all known offenses, mostly theft and burglary without violence.

E. Drug Testing Issues

Drug testing was performed by one of Holland's leading laboratories in this field, the Delta Laboratory of the Psychiatric Hospital in Poortugaal, near Rotterdam. The methods and standards applied in this project are equivalent to those of the U.S. ADAM program. Processing of the urine samples was done at the laboratory, usually within 3 days after the samples were obtained.

F. Instrument Issues

The standard U.S. ADAM interview list was translated into Dutch and adapted to match Dutch circumstances, (e.g., changes were made to the survey for the categories of housing, education, cultural background, and the like). Some questions were dropped at the request of the Bar Association, which believed that these questions would be too intrusive or that they could interfere with the judicial process. Among the questions dropped were those on how illegal income was obtained, the purchase of drugs shortly before arrest, and the time of arrest. Some new questions were added, such as reason for arrest, place and town district of residence, mode of use of drugs (e.g., injection or inhalation), use of medicines (some prescription drugs can cause false positives in testing), and more categories of nationality. Finally, the Dutch questionnaire was translated back into English and French. Interviews usually went smoothly, although for users of multiple drugs, the repetitive questions could become somewhat tedious.

Cultural applicability

With the adaptations mentioned above, the questionnaire proved to be quite user-friendly. Some interviews were held in other languages (English and French). Some respondents were not native speakers of any of these languages, and some of these were also illiterate; in these cases, explaining the questions and obtaining the right answers sometimes proved difficult.

Standardization versus local utility

Most of the material obtained in the pilot test can be compared with that collected in other countries, especially the United States. This applies to both the urinalysis and the encoded answers. Since the questionnaire used in the Dutch I–ADAM was adapted to fit standard categories used in other addiction programs and studies, local utility should be guaranteed as well.

In interpreting the comparative results, however, differences in legal and socioenvironmental conditions should be considered. Prison conditions, for example, may differ from country to country. Police stations in the Netherlands are spacious, clean, and relatively comfortable. Arrestees there may be less willing to cooperate in a research project than arrestees in countries such as the United States, where circumstances in police cells are less pleasant. Another difference among countries is whether the use of soft drugs is forbidden. In countries such as the Netherlands, Belgium (since 2001), and Switzerland, where the use of cannabis is not seen as a crime anymore, people obviously will be more willing to tell researchers that they have used cannabis. In countries such as the United States or Malaysia, where the use of soft drugs is seen as a crime, people are more reluctant to admit drug use, even in an anonymous research setting.

VII. Barriers to Developing I–ADAM

The Bar Association's objections resulted in an arrangement under which every candidate had to grant written permission to be interviewed. This introduced a formal element in the introduction procedure involving a routine (i.e., the signing of an informed consent form) that many people in police custody presumably have negative associations with. Several people could not read or write any of the four languages used. The need to hand out and collect informed consent forms added yet more work to the time-consuming and sometimes irritating formalities that were part of the I–ADAM procedure. The researchers believe that the obligation to obtain written consent from the arrestees resulted in a lower response rate. An additional problem is that, whereas normally, after the interviewing period, the research organization would not have any data that could ever be traced to individual persons, there is now a collection of 125 signed statements on which arrestees have quite often put their full name in legible writing. Under the agreement reached with the Bar Association, these statements have to be kept for a certain period of time in case any of the arrestees were to claim that he or she had been forced into participating in the project.

Other barriers were more technical in nature. Interviewers were not able to interview many potential candidates because they had been removed from the premises to be presented to the investigating magistrate very early in the morning after having been brought to the central holding facility at police headquarters late the previous night. In such cases, they remained out of reach for the interviewers, who could not visit them earlier in the day at a district police station or during the night, when the holding facility functions with only a limited number of guards. To cover all the potential candidates, researchers would have to interview at all of the more than 20 police stations in the region and at all hours at the central facility. This would pose substantial financial problems for both the research project and the police force, which would have to bring in additional guards if it were to grant interviewers access to the cells.

Another issue that emerged was the language barrier: Some of those approached for an interview spoke only Turkish, Moroccan (Berber), or an African or Eurasian language not spoken by the researchers. It would seem infeasible and impractical to prepare for interviewing in all possible languages.

VIII. National Stakeholders Involved in the Development of I–ADAM

As was mentioned before, all stakeholders whose approval was required for the pilot to be carried out were interviewed, and their permission was obtained in the fall of 1998. These stakeholders included the Ministry of Justice, WODC, the Public Prosecutions Department, the chief commissioner of the police force, the local police where the pilot tests took place, the Bar Association, and the Burgomaster of The Hague. Simultaneously, the relevant senior officials at the Ministry of Justice and the Procurator General responsible for narcotics issues on a national level in the Public Prosecutions Department were informed as well.

IX. Research Methods

To establish which subgroups would have to be represented in a selected arrestee sample, the characteristics of the overall population of arrestees in the Hague region had to be understood. Because such information is not available from police figures, there was no reason to draw up criteria for a weighted sample of candidates. Since the overall number of arrestees in any one Dutch police region is lower than that in typical U.S. ADAM sites, it was considered best simply to approach every arrestee who met the established criteria until the agreed number of interviews was obtained. Because this was a pilot project, as many persons as possible had to be tested in a relatively short time, so the easiest (rather than the most representative) sampling scheme was selected. Subjects in the pilot project had to meet the following criteria:

- No persons under the age of 16.
- No persons who had been arrested more than 48 hours before.
- Only persons arrested on suspicion of a crime.
- Only persons who understood and spoke Dutch, English, German, or French.
- No persons who were mentally unstable, incoherent, or extremely aggressive.
- No persons arrested for traffic violations.

X. Training Issues

The interviewers were junior researchers and students working for ES&E who received instructions on how to do the interview from senior ES&E researchers with extensive interviewing experience. No other persons (including law enforcement officers or court personnel) were allowed to participate in the interviews with the arrestees. In preparing for the instruction sessions, the I–ADAM training material was put to good use, but no extensive training exercise sessions were held to establish whether interviews observed by multiple interviewers would produce similar reports. During the interviews, interviewers were asked to keep a logbook to record every remarkable or unusual experience they encountered or observation they made. These were analyzed, but no substantial problems influencing the integrity or professionalism of the research work were recorded.

XI. Analysis of Interview Data

Although obtaining and analyzing the interview data was, in fact, of secondary importance in the initial pilot study, some results were worth mentioning. Readers have to keep in mind, however, that at this early stage, the small sample size (only 125 interviews) and the relatively low response rate can result only in indicative data that should be interpreted with the greatest caution. Any comparisons with I–ADAM data from other countries should perhaps be restricted to hypothesis-forming impressions and should not be used to form conclusions about the drug situation in the Netherlands. In this brief report, only a limited part of the results is presented.

A. Response Rate

Thirty-eight percent of the total number of arrestees fell outside the agreed category of arrested suspects taken into legal custody (such as suspects arrested and then released) (see exhibit 7). Of those that would qualify on this condition, another 20 percent could not be interviewed because they did not meet other criteria (such as age or length of time after arrest). Of those who met all the criteria, 23 percent were missed because they were either in transport or in interrogation and therefore could not be interviewed. Finally, 49 percent of those who met the criteria and were approached refused to cooperate.

Those who refused to cooperate gave the following reasons: An estimated 70 percent said that they saw no benefits in cooperation; about 20 percent objected to giving permission in writing (a condition required by the Bar Association); and about 10 percent were approached by a guard in a way that did not motivate them to cooperate. Of the 125 respondents (51 percent of those approached), 80 (64 percent) provided urine samples for analysis, 43 refused to give a sample, and 2 were incapable of urinating. A total of 111 men and 14 women were interviewed. The age breakdown of program participants is shown in exhibit 8.

Exhibit 7. Arrestees Participating in Dutch I–ADAM Pilot Program

Arrestees at the police station during the pilot	651
Not taken into legal custody:	248
Did not meet criteria (e.g., age)	82
Could not be interviewed (e.g., transport)	75
Total of available arrestees who met criteria	246 (38% of total population)
Willing to participate in interview	125 (51% of approached arrestees)
Gave urine samples	80 (64% of persons interviewed)
Were unwilling to give urine samples	43 (34% of persons interviewed)
Were unable to give urine samples	2 (2% of persons interviewed)

Exhibit 8. Age of Dutch I–ADAM Pilot Program Participants

Age (years)	Percent
16–20	24.8
21–30	40.8
31–40	25.6
41–50	7.2
51–65	1.6
Total	100.0

$N = 125$

B. Offenses

The grounds for arrest (category of primary offense) are presented in exhibit 9.

C. Reported Drug Use

None of those who declared themselves addicted reported currently being in any drug rehabilitation program, although many stated that they would like to participate in such a program. Respondents were asked if they had ever used drugs, and if they used drugs during the past 12 months and or during the past 3 days. The results are shown in exhibit 10.

Approximately 45 percent of the arrestees reported using alcohol in the 3 days before the interview, and 34 percent said that they used cannabis during that time. Percentages for other drugs are lower—approximately 13 percent for cocaine and opiates.

Exhibit 9. Offenses for Which I–ADAM Pilot Program Participants Were Arrested

Offense	Percent
Sexual offenses	1.6
Violent offenses (including murder and attempted murder)	32.0
Property crimes	46.4
Opium law	9.6
Other	8.0
Unknown	2.4

Exhibit 10. Self-Reported Drug Use of I–ADAM Program Participants

Drug	Ever used (%)	Used in the past 12 months (%)	Used in the past 3 days (%)
Alcohol	92.8	83.2	45.6
Cannabis	74.4	56.8	34.3
Crack	17.6	12.8	9.6
Cocaine	48.0	27.2	12.8
Opiates (mostly heroin)	24.8	19.2	13.6
Methadone	21.6	14.4	8.0
Amphetamines	16.8	4.0	0.8
Benzodiazepines	14.4	10.4	7.2
Other drugs	24.8	12.8	2.4

In total, 48 of the 125 respondents (38 percent) reported some form of addiction, and 26.4 percent have been treated for some kind of addiction at some time in the past. There are many "poly-users," mostly combinations of heroin and cocaine or crack, but also quite a few persons addicted to only one narcotic. Apart from the 48 self-declared addicts, about 10 people have a daily and often substantial drug intake but do not consider themselves addicted.

When the results of the urine tests are compared with self-reported use, a big difference can be seen in the results for alcohol. Although approximately 45 percent of the arrestees told researchers that they had used alcohol in the past 3 days, only 4 percent tested positive for alcohol in urine tests. This clearly shows that urinalysis is not an effective way to test for alcohol. For cannabis, cocaine, and opiates, however, urinalysis results showed a higher rate of use than did self-reports from arrestees. For cannabis, the difference is relatively small (34 percent of arrestees said that they used it, but 41 percent tested positive in the urinalysis). As expected, differences for cocaine and heroin were much larger. Approximately 13 percent of the arrestees said that they had used cocaine in the past 3 days, but urine tests showed 32 percent testing positive; 13 percent of the arrestees reported using heroin, whereas urine tests showed 17 percent testing positive. Apparently, arrestees are more honest about their use of cannabis. This is not a surprise, since the use of cannabis is not considered to be an offense.

When compared with the lifetime use of drugs in the country (exhibit 5), one can see that arrestees use drugs more frequently than the general population. Many of the arrestees in the sample were arrested for theft and violent offenses, and the relatively high rate of cocaine use may be related to this. Research shows that cannabis use in the Netherlands is not higher, and may even be lower, than in other countries, such as the United States. It will be interesting to compare the results for arrestees in these two countries.

D. Tested Drug Use

The results of the urine tests of the 80 persons who gave a urine sample are given in exhibit 11. Because many more respondents (28 percent) reported drinking alcohol shortly before being arrested than is reflected in the urine test, this test cannot be considered very useful. In view of the importance of the relationship between alcohol and crime, the administrators of the program will have to determine how to deal with this problem and whether alcohol can be measured in some other way.

Of the 80 arrestees willing to give a urine sample, 61 percent tested positive for some drug. Given the limited quantitative basis of 125 interviews, however, and the relatively low response rate on the urine sample, these data should be interpreted with the greatest caution.

E. Drug Use Before Committing the Offense

Respondents were asked whether they had used alcohol or drugs shortly before committing the offense. The results are shown in exhibit 12.

Those who reported that they used drugs shortly before committing the offense (44 percent) were asked which drugs they used. The results are shown in exhibit 13.

Exhibit 11. Urine Test Results

Drug	Positive urine sample	
	Number	Percent
Alcohol	3	3.8
Cannabis	33	41
Cocaine	26	32
Ecstasy	1	1
Opiates	14	17
Methadone	7	8
Benzodiazepines	13	16
Any drug	49	61

Exhibit 12. Alcohol and Drug Use Before Committing Offense

Drug	6 hours before offense (%)	Less than 12 hours before (%)	Less than 24 hours before (%)
Alcohol	28.0	5.6	2.4
Other drugs	36.8	5.6	1.6

$N = 125$

Exhibit 13. Type of Drug Used Before Committing Offense

Drug	Percent
Cannabis	43.6
Cocaine	25.5
Heroin	25.5
Crack	9.1
Ecstasy	3.6
Amphetamines	1.8
Methadone	1.8

Note: Some I–ADAM program participants reported use of more than one drug.

F. Offenses and Drug Use

The relationship between offenses and drugs found in the urine samples was examined. The results are shown in exhibit 14, an overview of the primary offenses (of which 46 percent were theft, 27 percent violent offenses, 2 percent sexual offenses, 13 percent opium, and 1 percent other offenses) of the 80 arrestees who were willing to give urine samples. From this overview, one can see that 19 of the 37 persons arrested for theft (51 percent) tested positive for cocaine. Persons arrested for theft and violent offenses test positive for a relatively large number of drugs. Persons arrested for

Exhibit 14. Offenses and Drug Use

Offense	Total	Alcohol	Cannabis	Cocaine	Ecstasy	Opiates	Methadone	Amphetamines	Benzo	No drugs found
Theft	37	3	19	19	1	9	6	1	9	7
Violent offenses	22	0	10	4	0	2	1	0	3	8
Sexual offenses	2	0	1	0	0	0	0	1	1	1
Drug offenses	11	0	1	2	0	1	0	0	0	9
Other offenses	8	0	2	1	0	2	0	0	0	5
Total	80	3	33	26	1	14	7	2	13	30

Note: Drug use determined using urine test.

other offenses test positive for relatively fewer drugs. Again, these data should be interpreted with the greatest caution, since the number of urine samples and the response rate were relatively low.

G. Previous Arrests

Of those interviewed, 41.6 percent reported having been arrested at least once in the past year; 55.8 percent of these had been arrested once before, the others more frequently. Going back even farther, 59.2 percent of all respondents reported having been arrested before the past year and 52.7 percent of all respondents reported having been detained in prison.

H. Cultural Origin and Nationality

More than three-quarters of all respondents (77.2 percent) claimed Dutch citizenship, although 57.7 percent were of a non-Dutch ethnic background (i.e., their parents were, or had originally been, citizens of other countries). Most of these were Moroccan, Surinamese, or Turkish.

XII. Policy Implications and Applications of Data

This I–ADAM pilot project has shown that research on arrestees may produce valuable information that could be useful for the further development of Dutch drug policy, both from the standpoint of maintaining public order and in terms of the treatment of drug addiction. The low response rate (51 percent), however, gives reason for concern. The experience of running the pilot project has given researchers some ideas on how to improve the response rate, but additional testing would be required to prove that a higher response will be possible. The relatively costly and time-consuming procedure is another cause for concern.

After the pilot program concluded, it was up to the Ministry of Justice to decide whether I–ADAM would be continued on a broader scale in the Netherlands. The report of the pilot program were presented to the persons involved from the Ministry of Justice and the Ministry of Health, Welfare, and Sports, and the results were discussed within the two ministries and with Dutch health care officials. If these persons are interested, the results will be brought to the attention of other interested parties (such as policy and care organizations) in other police regions in the Netherlands. A possible next step could then be to test I–ADAM in three other police regions to find out whether the response rate would be higher and whether more data would be received. After that, the Ministry of Justice could decide whether to expand the program further.

The results of the pilot study were made available at the end of 1999 and were discussed with policymakers at the Ministries of Justice and Health, Welfare, and Sport in 2000. The study report held that if the pilot program were to be continued, the response rate would have to be higher and arrestees would have to be tested for alcohol as well. If so, the project could provide interesting data.

Policymakers in these ministries, however, decided not to continue the project. One reason for this may be that much is already known about Dutch drug users. The main reason for not continuing the project was that policymakers did not see that response rates could be improved enough to generalize results to the whole population of arrestees. Much is already known about drug users in the Netherlands. Policymakers thus chose to give priority (and allocate funding) to other research, including research on a new system of compulsory treatment for the most persistent drug-related offenders (the "SOV" project), in which offenders who have committed numerous offenses are now forced to join a 2-year program, in which they spend the first 6 months in jail and then go to other surroundings, where they are helped to stop using drugs (or at least to use them in a more controlled way), and to find work and housing. This project will be evaluated extensively and will provide information about the group of arrestees.

Although SOV will give information about arrestees, the authors regret that I–ADAM will not be continued, especially because I–ADAM could have produced important international comparative data. The only chance for the ADAM project to be resumed in the Netherlands is if those running the program can work out a reliable way to test arrestees for alcohol, or, if it turns out that I–ADAM can generate information needed by parliament, it is possible that the project may be continued after all.

If it were expanded to more regions on a regular basis, the Dutch I–ADAM program promises to become an important information resource that would—

- Produce better and more policy-relevant information on drug use (quantities and sorts of narcotics, monitoring specific user groups and risk groups) and on the relationship between drug use (including alcohol) and specific forms of crime.
- Help to discover and confirm new drug use trends.
- Provide more and better information on local trends and relationships between drug use and crimes and eventually lead to the development of user/perpetrator profiles.
- Improve understanding of user groups that have so far escaped the focus of addiction and drug treatment programs, with the possibility of improving the range and effectiveness of narcotics care for criminal users.
- Supply the material needed for international comparative studies within the broader I–ADAM framework.

Additionally, if the questionnaire currently used can be expanded with some of the questions that had to be removed earlier at the request of the Bar Association, operationally relevant information on local drug markets and channels and ways of distribution would become available.

In summary, the results of the I–ADAM pilot in the Netherlands suggest that the program would improve understanding of drug use among people committing crimes and help shape future drug policy. The relatively low response rate and the difficulties in measuring alcohol in urine samples, however, are reasons for concern that cannot be easily overcome. In countries where response rates were higher, however, I–ADAM has the potential to provide new information about the relationship between drug use and crime. The authors believe that international data collection and comparison should be encouraged.

Notes

1. U.S. Census Bureau, International Database, 2000.

2. Department for Statistical Information and Policy Analysis, *Registered Crime in the Netherlands, 1995–1999*, The Hague: Ministry of Justice, 2000.

3. *National Drug Monitor: 2000 Annual Report,* Utrecht: Trimbos Institute, 2001: 24–25.

4. Coffee shops can best be described as cafés (which are *not* allowed to sell alcohol) in which soft drugs may be sold under the following conditions:

- No more than 5 grams per person may be sold in any one transaction.
- No hard drugs may be sold.
- Drugs may not be advertised.
- No selling to minors may take place.
- No more than 500 grams may be kept in stock.
- Coffee shops must not create a nuisance.

5. Information in this paragraph is based on Tak, P.J.P., *The Dutch Criminal Justice System: Organization and Operation,* The Hague: Ministry of Justice, Research and Documentation Center (WODC), 1999.

6. Arrestees participating in the I–ADAM pilot were all taken into custody.

7. Information in this paragraph is based on the following Dutch Government publications: *Drugs Policy in the Netherlands: Documentation 1,* The Hague, Ministry of Health, Welfare and Sport, 1997; *Drugs Policy in the Netherlands, Continuity and Change,* The Hague, Ministry of Foreign Affairs, Ministry Health, Welfare and Sport, Ministry of Justice, Ministry of the Interior, 1995; *Nederlandse bijdrage inzake de implementatie van het Gemeenschappelijk Optreden Drugs, rapportage 1999; Drugbeleid, voortgangsrapportage September 1997–September 1999,* The Hague: Ministry of Health, Welfare, and Sport, 1999.

8. For example: Korf, D.J., *Dutch Treat: Formal Control and Illicit Drug Use in the Netherlands,* Amsterdam: Thesis Publishers, 1995; and Grapendaal, M., E. Leuw, and H. Nelen, *A World of Opportunities: Life-Style and Economic Behavior of Heroin Addicts in Amsterdam,* Albany: State University of New York Press, 1995. Much information on drugs research can be found at the Netherlands Institute of Mental Health and Addiction (Trimbos-Institute) in Utrecht *(http://www.trimbos.nl).*

9. These numbers add up to more than 100 percent because they include offenders who committed one or more of the listed offenses.

SCOTLAND

Neil McKeganey
Clare Connelly
John Norrie
Janusz Knepil

Neil McKeganey is the director of the Centre for Drug Misuse Research at the University of Glasgow. **Clare Connelly** is a lecturer at the School of Law, University of Glasgow. **John Norrie** is the assistant director of the Robertson Centre for Biostatistics, University of Glasgow. **Janusz Knepil** is an honorary lecturer in the Department of Pathological Biochemistry at the University of Glasgow and a member of the Department of Biochemistry at Gartnavel General Hospital, Glasgow.

Introduction

This chapter describes research piloting the Arrestee Drug Abuse Monitoring (ADAM) Program in Scotland and covers the following background:

- Negotiation of access to Scottish police stations.
- Development of the research instrument.
- Drug testing procedures used in the pilot.
- Extent of drug abuse identified among arrestees.
- Practical difficulties of undertaking the pilot.
- Policy implications arising from the pilot and the next stage of ADAM research in Scotland.

I. Overview of Demographic Profile of Nation

Scotland is a devolved part of the United Kingdom. Although it shares some powers of government with the Parliament in London, Scotland has its own Parliament and distinctive legal system. Exhibits 1 and 2 present some basic demographic and economic information about Scotland.

Police in Scotland recorded 432,000 crimes in 1998, up from 421,000 in 1997. In 1999, 118 murders and 712 attempted murders were committed in Scotland; 6,323 serious assaults and 5,075 incidents of robbery and assault were reported to police.

Exhibit 1. General Demographic Information for Scotland

Geographic area	31,510 mi² (78,789 km²)
Estimated population	5.1 million
Gender distribution	50.8% female, 49.2% male (U.K.)
Age distribution	0–15 years, 19.9%; 16–64 years, 64.7%; ≥65 years, 15.4%
Life expectancy (1998)	Scotland, 72.1 years (male), 78.1 years (female); United Kingdom, 74.9 years (male), 79.8 years (female)
Population growth rate	–0.09%
Birth rate	10.56 per 1,000
Death rate	11.67 per 1,000
Languages	English, Scottish Gaelic (approx. 60,000 speakers)
Literacy rate	<99% (1978 est.)
Political subdivisions	9 regions (Borders, Central, Dumfries and Galloway, Fife, Grampian, Highland, Lothian, Strathclyde, and Tayside); 3 island areas (Orkney, Shetland, and Western Isles)

Sources: Registrar General for Scotland, *Annual Report of the Registrar General for Scotland, 1999*. Edinburgh, General Register Office for Scotland, 2000; Central Intelligence Agency, *The World Factbook: United Kingdom*, Washington, DC: Central Intelligence Agency, 2000.

Exhibit 2. Economic Information for Scotland

Gross domestic product (GDP) (1997)	£56.2 billion
Average weekly earnings	Scotland, £394.6 (males), £276.7 (females); United Kingdom, £425.6 (males), £308.7 (females)
Unemployment rate	7.1%

Source: Scottish Executive

Although little information exists on the size of the problematic drug-misusing population within Scotland, it is clear that the country has a substantial and growing drug problem. A comparison of the results of national household surveys conducted in 1993 and 1996[1] shows that self-reported lifetime drug use increased for all drugs and use in the previous 12 months increased for all drugs except LSD (see exhibit 3). In 1996, 22.5 percent of the Scottish population had ever used illegal drugs, compared to 9 percent in the previous year. Cannabis was by far the most widely used illegal drug, followed by amphetamines and LSD. The drugs with the largest proportionate increases in use were temazepam (benzodiazepene) and Ecstasy (MDMA); lifetime use rates for both doubled from 1993 to 1996 (see exhibit 3). On the basis of the 1996 Scottish Crime Survey, approximately 679,000 people within Scotland have tried at least one illegal drug, 560,000 have used cannabis, and 41,000 have used opiates.[2]

Research has identified high levels of drug misuse among young people in Scotland. By age 15, 40 to 50 percent of young people have started to experiment with illegal drugs.[3] Recent research has also shown that young people in Scotland are starting to use illegal drugs at an increasingly young age; McKeganey and Norrie have reported that 10 percent of 11- to 12-year-olds in Scotland have already begun to use illegal drugs.[4]

Exhibit 3. Prevalence of Self-Reported Drug Use in Scotland, 1993 and 1996, Ages 16–59

	1996		1993	
	Lifetime use (%)	Use in past 12 months (%)	Lifetime use (%)	Use in past 12 months (%)
Any drug	22.5	9.0	18.3	6.8
Cannabis (marijuana)	19.0	8.2	14.5	6.1
Amphetamines	7.3	3.1	5.2	2.2
LSD	5.5	1.3	4.0	1.6
Psilocybin (mushrooms)	5.1	1.3	3.7	1.0
Ecstasy (MDMA)	4.0	2.1	2.0	1.1
Temazepam (benzodiazepine)	3.7	1.1	1.7	0.9
Valium	4.2	0.9	2.7	0.3
Solvents	2.1	0.2	1.6	0.1
Powder cocaine	2.6	1.0	1.5	0.4
Crack cocaine	0.7	0.2	0.4	—
Heroin	0.8	0.1	0.4	—
Methadone	0.9	0.3	0.2	0.1

Source: Anderson, S., and M. Frischer, "Drug Misuse in Scotland: Findings From the 1993 and 1996 Scottish Crime Surveys," *Crime and Criminal Justice Research Findings*, No. 17, Edinburgh: The Scottish Office, Central Research Unit, 1997.

The rates of injection drug use, drug-related mortality, and hepatitis C and HIV infection among injection drug users (IDU) are sources of concern in Scotland. Glasgow, Scotland's largest city, has an estimated 9,500 injecting drug users (IDU).[5] Dundee, Scotland's third-largest city, has an estimated 3,500 opiate and benzodiazepine misusers.[6] Drug-related mortality has been increasing in Scotland over the past several years.[7] An estimated 60 percent of IDUs in Scotland may be hepatitis C positive.[8] Approximately 20 percent of IDUs in Edinburgh (Scotland's capital and second-largest city) and 26.8 percent of IDUs in Dundee are thought to be HIV positive.[9] In Glasgow, however, only 1 to 2 percent of IDUs are thought to be HIV positive.[10]

Relatively little research has been undertaken to quantify drug-related crime within Scotland. Recent research has estimated that IDUs in Glasgow spend approximately £94 million each year on drugs. To support that level of drug purchasing, an estimated £129 to £258 million worth of goods is being stolen each year.[11]

II. Description of Criminal Justice System, Arrest Process, and Detention System

Police in Scotland can make arrests when they witness, or are called to the scene of, a crime and the accused is present. In these circumstances, the suspect is often immediately charged with an offense and either detained in custody in the police station, to appear in court the next morning, or set free on an undertaking to appear in court at a future date. If a warrant is issued for an accused's arrest, the accused is detained in the police station and appears in court the next day. If an offense is committed over the weekend, the accused is held in custody until Monday morning, when the first court sits. Courts in Scotland do not sit on Saturday or Sunday.

Following an arrest or a charge, a report is sent to the procurator fiscal, who decides whether a prosecution should take place and, if so, what the charges should be. Procurator fiscals are civil servants who prosecute crimes in the lower courts in Scotland and draw up charges for the higher courts, where a case is prosecuted by an advocate depute. The case is prosecuted if the procurator fiscal decides that sufficient evidence exists and that prosecution is in the public interest. The procurator fiscal decides whether or not there should be a jury trial; an accused has no right to demand a jury. In cases of rape and murder, however, trial must be by jury.

An arrestee held in police custody is given a statement of the charges on which he or she is to be prosecuted in the court appearance the next day. At this hearing the arrestee is either remanded to custody (prison) to await trial or set free, pending trial. If the arrestee is remanded to custody, the trial must start within 110 days; if the arrestee is set free, the trial must start within 1 year of the date of appearance. If the prosecutor fails to bring the case to trial in a timely fashion, the accused is set free, and no future trial on the same charges can occur. If not detained in police custody, an arrestee is set free on an undertaking to attend a specified court at a specified time.

Scottish criminal law is mainly a body of common law; hence, limited criminal statutes and no criminal code exist. For all common law crimes except murder, sentence following conviction is at the judge's discretion. Murder carries a mandatory life sentence, with release being on life license. For statutory offenses (e.g., all drug offenses within the Misuse of Drugs Act, 1971), judges' sentencing powers are detailed in the legislation.

III. Status of National Drug Policy

The primary U.K. legislation relating to drug abuse is the Misuse of Drugs Act, 1971. Section 1 of the Act authorizes the creation of the Advisory Council on the Misuse of Drugs, a group of Government officials and external experts who advise the Government on the changing nature of the drug problem in the United Kingdom. Section 2 of the Act identifies the range of drugs falling within the terms of the Act and places these in one of three classes. Exhibit 4 sets forth the drugs in each class and the penalties for possession for personal use and possession with intent to supply.

Government anti-drug-abuse policy is set forth in the documents, *Tackling Drugs to Build a Better Britain*[12] and *Tackling Drugs in Scotland: Action in Partnership*.[13] Both documents present Government goals under four key headings: young people, treatment, communities, and availability (see "Objectives of Scottish Drug Misuse Strategy").

IV. History of I–ADAM in Scotland

The idea of conducting an ADAM study within Scotland came from John Hamilton, Chief Constable of Fife Constabulary. Chief Hamilton had attended the first annual I–ADAM meeting

Exhibit 4. Classes and Penalties of Drugs in the United Kingdom

		Maximum penalties*	
Class of drugs	Types of drugs	Possession for personal use	Possession with intent to supply
A	Cocaine including crack, opiates (heroin, morphine, opium), LSD, methadone, Ecstasy (MDMA), injectable amphetamines, PCP, magic mushrooms, cannabinol and derivatives	7 years' imprisonment	life imprisonment
B	Cannabis (marijuana, herbal and resin), barbiturates, noninjectable amphetamines, codeine, dihydrocodeine, methylamphetamines	5 years' imprisonment	14 years' imprisonment
C	Benzodiazepines (temazepam, Valium), anabolic steroids	2 years' imprisonment	5 years' imprisonment

*All penalties are in addition to or in place of an unlimited fine.

Source: Independent Inquiry Into the Misuse of Drugs Act, 1971. *Drugs and the law: Report of the Independent Inquiry Into the Misuse of Drugs Act, 1971,* London: Police Foundation 1999. [cited herein as Runciman Report 1999].

in Miami in April 1998 and saw the value of applying the ADAM methodology in Scotland. Following discussions with government officials, funding was provided for a pilot of the ADAM approach in two police jurisdictions in Scotland.

Shortly after the first annual I–ADAM meeting, National Institute of Justice (NIJ) staff from the U.S. Department of Justice visited Scotland in June 1998 to provide government officials with details of ADAM operations in the United States and gain a better understanding of how ADAM would operate in Scotland. In August 1998, Scottish officials conducted a press briefing with the Fife police force regarding their work in developing an I–ADAM site in Fife and the process for selecting a research team. In October 1998, the Scottish Office conducted a second press briefing to announce the inclusion of the Strathclyde Police Force in the I–ADAM project.

The pilot was carried out over 6 months from June to November 1999.[14] Both police forces were actively involved at all stages of the research and both sent representatives, along with the research team, to the 1999 and 2000 I–ADAM annual meetings.

V. Goals and Objectives of I–ADAM in Scotland

The Scottish ADAM pilot aimed to—

- Test the feasibility of using the ADAM methodology within the Scottish context.
- Consider the requirements of ADAM data collection sites and their application in Scotland.
- Consider the value of using the ADAM methodology to monitor and assess changes in drug use over time.
- Gather information about the type and extent of drug use among recent arrestees in the selected areas.
- Compare self-reported drug use with the results of urine testing.

- Consider the ways in which arrestees describe and justify their behavior.
- Assess the utility of additional questionnaires for identifying and exploring issues potentially related to drug use.

VI. Getting Started

Although there was considerable support for an ADAM pilot at the planning stage, all key stakeholders involved in the research (police, officials from the Crown Office, and senior civil servants from the Health and Criminal Justice Departments) were acutely aware of the problems and sensitivities that the study was likely to involve. Never before in Scotland had an independent

---

Objectives of Scottish Drug Misuse Strategy

Strategic Aim 1: Young people—to help young people resist drug misuse in order to achieve their full potential in society.

Key Objective: Reduce the proportion of young people under age 25 reporting use of illegal drugs in the past month and the previous year.

- Establish a consistent and coordinated evidence-based approach to drug education, prevention, and harm reduction that takes account of individual and community needs.
- Implement education strategies and initiatives and provide public education that increases knowledge and awareness of drug misuse and promotes the avoidance of drug misuse.
- Reduce the acceptability and incidence of drug misuse among children and young people.
- Ensure that every school pupil in Scotland has effective drug education, including accurate and up-to-date information on the consequences of drug misuse.
- Reduce the harm from drug misuse by encouraging positive alternative lifestyles.
- Increase access to services for vulnerable young people, including school excludees, truants, looked-after children, young offenders, young homeless, very young children at risk of drug misuse, and the children of drug-using parents.

Strategic Aim 2: Communities—to protect our communities from antisocial and criminal behavior.

Key Objective: Reduce levels of repeat offending among drug-misusing offenders.

- Strengthen and protect communities from drug-related crime and the fear of drug-related crime.
- Reduce the level of drug misuse in prisons.
- Develop constructive alternatives to prosecution and imprisonment for drug-related offenses.
- Support partnerships among professionals, local people, and businesses in developing local initiatives for tackling drug misuse.
- Promote drug awareness and the development of drug policies and health promotion in the workplace.
- Ensure that drug misuse is addressed within the wider context of area regeneration and social inclusion.

agency undertaken research interviewing individuals within hours of their arrest. The concerns voiced were of several kinds, including—

- Human rights concerns:

  — How to ensure that arrestees provide free and full informed consent to take part in the study.

  — How to ensure that no one participating in the study would be held in custody for longer than would otherwise occur.

- Practical concerns:

  — Whether to interview arrestees in the cells or in an adjacent area.

  — How to ensure that no other arrestees could overhear answers to specific questions.

Objectives of Scottish Drug Misuse Strategy (continued)

Strategic Aim 3: Treatment—to enable people with drug problems to overcome them and live healthy and crime-free lives.

Key Objective: Increase participation of problem drug misusers, including prisoners, in drug treatment programs that have a positive impact on health and crime.

- Reduce the health risks to individuals and communities from drug misuse, and reduce related infectious diseases.

- Increase the number of drug misusers becoming and remaining drug free, and promote their inclusion in society.

- Reduce the incidence of injecting, sharing, and polydrug misuse among drug misusers.

- Reduce the number of drug-related deaths.

- Increase the proportion of drug misusers in contact with services, including those in prison, through the development of good, accessible, responsive, and effective services.

- Reduce the number of drug misusers who have no quick access to appropriate treatment.

Strategic Aim 4: Availability—to stifle the availability of illegal drugs on our streets.

Key Objective: Reduce access to drugs among 5- to 16-year-olds.

- Reduce access to drugs among all age groups.

- Enforcement agencies to continue to identify, investigate, and obtain evidence for the prosecution of persons engaged in drug trafficking and supply.

- Reduce the amount of drugs entering Scotland by targeting distribution networks and dismantling organized trafficking.

- Support local constructive community efforts at reducing the local availability of drugs.

- Reduce the availability of drugs in prisons.

Source: Scottish Office, *Tackling Drugs in Scotland: Action in Partnership*, Edinburgh: The Scottish Office, 1999.

- Interviewer safety concerns:

  — What kind of arrangements would be needed to reduce the possibility of interviewers being assaulted during face-to-face interviews with arrestees.

- Concerns with obtaining and analyzing urine samples:

  — What kind of collection devices to use to obtain urine samples.

  — How to avoid possible risks of injury from sharp edges on collection devices, and how to keep collection materials from being bent or manipulated to create a sharp edge for use as a possible weapon.

- Confidentiality issues:

  — What level of confidentiality to afford the interviewee who admits to a serious crime.

A number of key stakeholder agencies, including the police, were concerned as to how an ADAM interviewer should respond if an arrestee admitted to having recently carried out a serious offense (e.g, murder) of which the police were unaware. Police and government officials believed that such information should be passed on to the police, but such a stipulation was problematic from a research standpoint because it would provide arrestees with only limited confidentiality when taking part in the interviews.

In the light of these and similar issues, a set of protocols was prepared that specified the procedures to be followed at each stage of the research. Topics covered by the protocols included the following:

- Selecting arrestees for interviews.
- Developing sample inclusion criteria.
- Scheduling interviews in the police stations.
- Recruiting arrestees into the pilot.
- Obtaining informed consent.
- Collecting and analyzing urine samples.
- Responding to inappropriate arrestee behavior during interviews.
- Recording information on all arrestees with participating custody areas over the period of the ADAM pilot.

Once the protocols had been agreed to by the commissioning group (which contained representatives of the Crown Office, senior civil servants from the Health and Criminal Justice Departments, and representatives of both police forces), the research was authorized to proceed and the research instrument was designed. After the team of interviewers for the research had been recruited and trained, letters were sent to all defense solicitors operating in the pilot areas informing them about the nature of the research that would soon be under way. Only two responses were received to the letters sent out from the research team: one congratulated the team on the valuable research they were doing, and the other requested a copy of the questionnaire and stated that the firm would be advising its clients to refuse to collaborate with the research. As it turned out, no clients of this particular law firm were knowingly involved in the pilot.

Facility Access

The ADAM pilot was carried out in Strathclyde and Fife. Strathclyde represents the largest area within Western Scotland and contains Glasgow, Scotland's largest city and home to some three-quarters of a million people. Fife, by contrast, is a predominantly rural area with small towns in eastern Scotland. Both Strathclyde Police and Fife Constabulary were actively involved at all stages of the research, including identifying the police stations where the ADAM program would be undertaken. At the outset, these were identified on the basis of the number of arrestees received. Two out of 115 police stations were identified in Strathclyde. All three of Fife's police stations were identified, but, in practice, the pilot was confined to the two main stations. The small number of arrestees processed in the third station would have required the fieldwork to be extended over a very considerable time period in order to recruit arrestees in sufficient numbers.

Both police forces requested an indemnity against possible legal action in the event of an interviewer being attacked by an arrestee in the course of this pilot. The University of Glasgow granted this indemnity on the condition that, when interviews were taking place, a police officer or force support officer would be positioned to see what was happening at all times and able to speedily intervene if an arrestee behaved aggressively toward an interviewer.

Once the participating police stations were identified, it was necessary to meet with the station commanders and their senior management staff to outline the nature of the proposed research. Each shift commander was briefed on the research procedures. In addition, police liaison officers associated with this research prepared a set of written protocols for custody officers to follow.

It would be difficult to overestimate the importance of this phase of the research. Clearly, the pilot could not have been conducted without the support of senior officers in both areas. Even with such support, however, this study also required the active support of all custody area staff who would be working directly with the interviewing team. Securing the involvement of this staff was not a once-and-for-all process. The ADAM pilot involved a continuous balancing between the needs of the research and the day-to-day demands of the police officers involved in the pilot. The following field note illustrates how the normal work of the custody area could affect the ADAM fieldwork:

> When Inspector Adams and I were going through the Arrestee Monitoring Forms (which provide details on all individuals arrested during each shift), we noticed that a substantial number of individuals had been arrested but had apparently not been considered for interview. When Inspector Adams spoke to the shift concerned, it was explained that an operation was under way within the station to target shoplifters and that, although a large number of shoplifters had been arrested, they had been processed very quickly and then released, leaving no time for them to be approached about a possible interview. Clearly, from a research point of view, it was disappointing to see so many arrestees being lost to interview, but to have held onto the individuals for the purposes of the research would have been unethical. Within such circumstances, there was nothing that could be done.

Drug Testing Issues

The key drug testing issue faced in the Scottish ADAM pilot was whether to rely on generic testing procedures or to undertake further confirmatory analyses of positive urine samples. It was decided to combine generic and confirmatory testing in the Scottish pilot for two reasons. First, because the study was a pilot, it was important to identify whether the confirmatory analysis would be necessary in any wider extension of the ADAM program in Scotland. Second, a variety of medicines containing low levels of opiates can be legally bought within Scotland; ingestion of these medicines can produce a positive result under generic urinalysis. Only by undertaking further confirmatory work could possible innocent opiate use be differentiated from more problematic forms of opiate misuse.

Clearly, one cannot, in this short report, discuss at length the results of generic and confirmatory drug testing analyses. Of the 87 urine samples that tested positive for opiates on the basis of the generic screening, four were reclassified as negative on the basis of the confirmatory analysis. Of the 21 urine samples that tested positive for amphetamines, no samples were reclassified on the basis of the confirmatory analysis. Two samples, however, that had been categorized as opiate positive but amphetamine negative were found, on further testing, to be amphetamine positive. One sample that had been classified as opiate negative but amphetamine positive was found to be opiate positive. These results are interesting but not conclusive, and they may indicate that combined generic and confirmatory analyses of urine samples are necessary.

Instrument Issues

The questionnaire used in this pilot was similar to that used in the first U.K. pilot (the New England and Wales Arrestee Drug Abuse Monitoring [NEW–ADAM] pilot). A copy of that questionnaire had been made available to the Scottish research team. In particular, the questionnaire covered the following areas:

- Demographic characteristics—age, education, employment.
- Legal and illegal drug use over the previous 12 months, 30 days, and 3 days.
- Drug buying activities.
- Arrestee's views of a possible link between his or her drug use and offending behavior.
- Extent of drug use and needle and syringe sharing.
- Personal expenditure on drugs.
- Sources and amounts of legal and illegal income.
- Past offending behavior.
- Use of drug treatment facilities and self-perceived need for help.
- Access to and carrying of weapons.
- Involvement in and experiences of domestic violence.

In addition to the standardized survey instrument, an arrestee monitoring form was designed and used for this pilot. This form, completed by the custody staff, recorded details on all individuals arrested during the pilot period. Through this means, the sample of arrestees who were interviewed could be compared with the broader population of arrestees in each custody area.

VII. Barriers to Developing I–ADAM in Scotland

There were no barriers as such to conducting the ADAM pilot in Scotland; however, the research raised many problems and difficulties that needed to be resolved in the execution of the pilot. At one level the ADAM methodology is very simple; indeed, what could be simpler than interviewing and drug testing members of a captive population? In fact, the research involves a complex array of problems, many of which could not be anticipated. Examples of issues that surfaced in the course of the pilot follow.

Security Cover

As stated earlier, the University of Glasgow granted an indemnity to police forces on the basis that, at all times when an interview was taking place, a police officer or force support officer would be in the vicinity of the interview, able to see what was happening but not able to overhear what was being said. An unanticipated consequence of this arrangement was that, during periods when the police stations were busy processing large numbers of arrestees, it was often impossible to release an officer to provide the necessary security cover. Conversely, when the custody areas were not particularly busy, it was easy to identify an officer who could provide security cover, but few arrestees were available for interview. The field note below illustrates this problem:

> On my usual trip to the station this afternoon, I spoke to one of the interviewers, who was sitting rather glumly in the room that had been allocated to us for the interviews. He said that there was a football match on in the area, that there had been considerable crowd trouble, and whilst a number of arrests had been made, the officer in charge of the shift had said that his staff were simply too busy to spare somebody to provide security cover to enable any interviewing to take place. When I then spoke to the officer in charge, he said that whilst he was sympathetic to the needs of the study, there was simply nothing he could do because all of his staff were fully committed to policing the football match. I then spoke to the interviewer and said that, although it looked unlikely that any interviewing would take place on his shift, he would still have to remain in the custody area in case the situation changed. When I asked if he wanted me to pick up a newspaper for him, the enthusiasm with which he greeted that suggestion spoke volumes about the prospect of another 4 hours sitting on his own in this tiny office.

The research often involved a complex negotiation with custody staff as to what was possible. Sometimes potential blockages in the interviewing could be resolved by reference to higher authority (a more senior officer); at other times, the interviewer simply had to recognize that other events occurring within the custody area were going to take precedence over the research. At such times, it was necessary to take a step back and wait for a more opportune moment when the interviewing could resume.

Confidentiality

At an early stage in this fieldwork, both police forces expressed the concern that, if an arrestee disclosed to the interviewer that he or she had committed a serious crime—for example, a murder or a kidnaping, this information would have to be passed on to the police as soon as possible at the conclusion of the interview. In effect, then, arrestees could be provided with only limited confidentiality. Clearly, it was essential that this limited confidentiality be explained to all arrestees. The research team agreed to provide arrestees with both written and verbal statements of the limited basis on which confidentiality could be granted to them in connection with their involvement in the ADAM pilot. This meant that arrestees' answers to questions covered on the research instrument could be considered confidential; however, any additional information volunteered by arrestees was not covered by the confidentiality clause. It was further explained to the arrestees that unsolicited disclosures would be regarded as inappropriate behavior and would result in the termination of the interview. In the planning stage of this research, this arrangement was considered to be a way of ensuring that the research team would not find themselves in the position of passing information to the police that had arisen in the course of an interview thought to be confidential by arrestees. As the following field note illustrates, however, even this arrangement was less than wholly adequate:

> At around 6 a.m., I received a telephone call from one of the interviewers who was clearly in an agitated state over an interview she had just completed. I said I would drive down to the custody area immediately, and, when I met her, she explained that midway through an interview, the arrestee had stated that his stepfather had abused him over many years. This was not an area covered by the questionnaire; the interviewer asked the arrestee to confine himself only to the questions that would be put to him and repeated that any additional information could not be regarded as confidential. The arrestee said that he understood this and wanted to continue with the interview and would stick only to answering the questions that were put to him. However, when the interviewer came to the section of the questionnaire that asks about the arrestee's access to guns, he indicated that he did indeed have access to a gun and that it was his intention, on being released from the police station, to kill his stepfather. The interviewer had then become very concerned about this disclosure. Unsure of whether she should have alerted the police to this threat, she had telephoned me. I was also unsure of how to respond to this occurrence. Since murder had not taken place the incident, as such, was not of the same kind as had been anticipated in the planning stage for the pilot. However, a threat to murder someone made within the context of a police station by an arrestee who has admitted access to a gun hardly seemed the kind of thing that one could simply ignore. Following a more general discussion with our police liaison officer and discussions with a colleague within the University, I made the decision that we had to alert the police to the nature of the threat that had been made. Since only one arrestee had been interviewed on that shift, it was not difficult to indicate to the relevant police officer which arrestee had made the threat.

It is important to recognize that, in responding to this incident, no clear guidelines existed as to how to proceed. As a result, the interviewer and her supervisor had to determine a response largely on the basis of what seemed right at the time. At the simplest level, it would have been intolerable to have been warned about a murder in advance of its occurrence and done nothing to prevent it. Since the likelihood that the individual would act in accordance with this threat could not be assessed, the only sensible option was to pass the information on to the police authorities.

VIII. National Stakeholders Involved in the Development of I–ADAM

Numerous parties have been involved in the development of I–ADAM in Scotland. These include the Scottish Executive, which funded this pilot program through a grant; the Fife Constabulary and the Strathclyde Police, which were the local police forces who cooperated in the program; the University of Glasgow, which sponsored the study and provided the interview team; and the Scottish Advisory Committee on Drug Misuse, which is the group responsible for setting and reviewing Scottish drug policy.

IX. Research Methods

The ADAM pilot in Scotland entailed a combination of research methods. First, custody staff were asked to complete an Arrestee Monitoring Form on all arrestees passing through each of the custody areas participating in this research. The form recorded key information to allow comparison of the arrestees who were interviewed and had their urine tested against the broader arrestee population within the study areas. Second, a sample of arrestees was interviewed about their drug use, using a standardized instrument (questionnaire). Third, the interviewees' self-reports of their drug use were confirmed by a laboratory examination of urine samples provided by those interviewed who consented to a urine test.

Sampling

The original plan was to have an interviewer available in each of the custody areas over the full 24 hours of each day of data collection. As a result of staffing difficulties, it was necessary to revise the interview schedules so as to make interviewers available 18 hours a day, 7 days a week. For each day of data collection, interviewer shift times were constructed so as to minimize the loss of any arrestees from this study as a result of the reduced coverage. The shift times used in this research were from 11 a.m. to 8 p.m. and from 9 p.m. to 6 a.m. in Strathclyde, and from 7 a.m. to 4 p.m. and from 6 p.m. to 3 a.m. in Fife. It became evident that most arrestees were in the custody area for longer than 2 to 3 hours, which meant that even if someone had been arrested in the period when no interviewer was present, the interviewer could usually request an interview when he or she came on shift.

Eligibility Criteria

The Scottish ADAM pilot included adult arrestees (age 16 and older) who could provide informed consent to participate in the research. The following arrestees were excluded from the pilot:

- Arrestees younger than 16.
- Arrestees who could not provide fully informed consent to participate in the research. These included arrestees who were impaired by alcohol or drug use, and individuals arrested on suspicion of murder. These last were excluded on the advice of the Crown Office because they are routinely assessed by a psychiatrist and, therefore, could not be assumed to be of sufficiently sound mind to provide fully informed consent to the research.
- Arrestees who the custody officer determined posed a threat of violence to the interviewer or whose interview might interfere with an ongoing police investigation.
- Arrestees in connection with drunk driving charges. Drunk drivers tend to be processed very quickly through police stations, and their recent alcohol use might impair their ability to provide informed consent to participate in the research.

Over the period of data collection in the Scottish ADAM pilot, 581 individuals were deemed ineligible for inclusion within the pilot. Exhibit 5 shows the reasons for the exclusion.

X. Training

A team of eight interviewers was recruited for this pilot, all of whom were either current or recently graduated students at the University of Glasgow. It was felt that using young people in this capacity would help in building up a rapport with arrestees and would counter any suggestion that the interviewers were undercover police officers. All interviewers were trained to use the research instrument and underwent rigorous security checks by the two police forces

Exhibit 5. Reasons for Ineligibility for I–ADAM Pilot Interviews

Reasons for ineligibility	Percent ineligible*
Impairment due to alcohol intoxication	26
Detained rather than formally arrested	23
Risk of potential violence by the arrestee	17
Arrestee younger than age 16	12
In custody too short a time	12
Drunk driving charge	8
Discretion of the custody staff	8
Impairment due to mental disorder	6
Impairment due to recent drug use	4

*Percentages add up to more than 100 because some arrestees were ineligible for more than one reason.

involved in this pilot. The police requested the security checks, given that the interviewers would be spending a considerable amount of time within secure areas of the police stations chosen as pilot sites.

The interviewing team comprised four men and four women. At the time the interviewers were selected, there was no way of knowing how well they would relate to arrestees or how close or distant they would be from the social background of arrestees. As it turned out, the vast majority of arrestees were from the poorer classes, while the interviewers generally had more middle-class backgrounds. More important than class, however, was the interviewers' ability to adopt a non-threatening stance toward arrestees and to reassure them that they were not undercover police officers. All interviewers wore visible identity cards noting their employment at the University of Glasgow.

XI. National Data Analytic Results

For this pilot program, 427 arrestees were interviewed, 220 in Strathclyde and 207 in Fife. In addition, urine samples were successfully collected from 67 percent (280) of interviewed arrestees. See exhibit 6 for demographic information on arrestees.

Exhibit 7 shows the results of urine testing in the Scottish ADAM pilot. Overall, 71 percent of urine samples were found to be positive for any drugs (excluding alcohol and volatile substances), as compared with 61 percent in the NEW-ADAM pilot.[15] Of the 280 arrestees who provided usable urine samples, 52 percent tested positive for cannabis. Cannabis use for male arrestees was 54 percent, compared to 42 percent for female arrestees. More than 50 percent of male arrestees tested positive for cannabis in only 1 of 35 U.S. ADAM sites (53.1 percent in Oklahoma); and more than 30 percent of female arrestees tested positive for cannabis in only two U.S. ADAM sites (31.9 percent in St Louis and 31.2 percent in Indianapolis).[16] These results indicate that cannabis use appears to be more prevalent among male and female arrestees in Scotland than among their counterparts in the United States.

Exhibit 6. Demographic Information About Scottish I–ADAM Arrestees (*n* = 427)

Pilot location arrested	Strathclyde, 220; Fife, 207
Sex	Male, 351; female, 75
Race	White, 100%
Marital status	Men, 67% single; women, 67% single
Employment status	Employed, 28%; unemployed/sick or disability benefit, 68%
Homeless status	16% within the past year
Education	38% ended at or before age 15

Exhibit 7. Results of Urine Testing (*n* = 280)

Drugs	Percent testing positive		
	All	Male	Female
All drugs	71	—	—
Cannabis (marijuana)	52	54	42
Benzodiazepines	33	—	—
Opiates ($p = 0.0009$)	31	26	51
Methadone ($p = 0.096$)	12	10	19
Amphetamines	8	—	—
Cocaine	3	—	—

— = Not available.

Thirty-one percent of urine samples in the Scottish ADAM pilot tested positive for opiates (29 percent in Fife and 33 percent in Strathclyde). Opiate misuse was more common among female arrestees (51 percent) than male arrestees (26 percent) ($p = 0.0009$). Methadone use rates were also twice as high among female as male arrestees in Scotland (see exhibit 7). More than 10 percent of adult male arrestees tested positive for opiates in only 8 of the 35 U.S. ADAM sites; none of the U.S. sites identified levels of opiate misuse greater than 20 percent among adult males. Twenty percent or more of female arrestees tested positive for opiates in only four U.S. ADAM sites (the highest figure was in Chicago, where 27 percent of female arrestees tested positive).

Exhibit 8 shows the results of the ADAM pilot interviews. In interviews, 89 percent of arrestees in the Scottish ADAM pilot reported having used an illegal drug at some time, 80 percent within the last 12 months. The self-reports confirmed the data from urine testing that cannabis was by far the most widely used illegal drug among arrestees; 71 percent reported having used cannabis within the past 12 months. Arrestee self-reports also confirmed the high prevalence of injection drug use in Scotland and of needle sharing among injection drug users. Twenty-six percent of arrestees reported that they had injected drugs; 22 percent had done so within the past 12 months. Of those who had injected drugs (110), 20 percent had used another individual's injecting equipment within the past 12 months, and 33 percent had passed on their own injecting equipment for another's use.

Exhibit 9 reports arrestees' self-reports of their own drug dependence, their perceptions of the links between drug use and crime and drug use and violence, and their access to weapons. These reports confirm the high rates of drug misuse among offenders in Scotland: 52 percent of arrestees stated that they felt they had been dependent on a drug other than alcohol at some point in the past, and 40 percent said that they had recently been drug dependent. IDUs were more than twice as likely as other arrestees to perceive a connection between their drug use and their offending behavior (84 percent for IDUs compared with 32 percent for all arrestees) and to own or have access to a gun (56 percent compared with 25 percent for all arrestees). Arrestees also strongly perceived links between drugs, especially alcohol, and violence (see exhibit 9). A significant minority of arrestees (20 percent) had carried a knife when committing a crime.

Exhibit 8. Self-Reports of Drug Use by Scottish ADAM Interviewees

Self-reports	Lifetime use (%)	Use in past 12 months (%)
Any illegal drug use	89	80
Dependence on illegal drugs (excluding alcohol)	52	40
Cannabis (marijuana)	—	71
Amphetamines	—	43
Heroin	—	31
Injection drug use	26	22
Use of others' injecting equipment	—	20*
Passing along injecting equipment to others	—	33*

*Percentage of those who reported injecting drugs within the previous 12 months.

— = Not available.

Exhibit 9. Self-Reports of Connections Between Drugs and Crime

Self-reports	Percentage reporting
Lifetime drug dependence	52
Recent drug dependence	40
Perceived connection between drug abuse and offending behavior	32
Perceived connection between drug abuse and offending behavior (current IDUs)*	84
Perceived connection between alcohol and violence (all arrestees)	93
Perceived connection between illegal drugs and violence (all arrestees)	65
Own or have access to gun (all arrestees)	25
Own or have access to gun (current IDUs)	56
Have carried knife when committing offense (all arrestees)	20

*IDU=Injecting drug users.

Exhibit 10 shows arrestees' self-reports of receiving treatment for drugs and alcohol misuse. The percentage of arrestees who stated that they would like to receive treatment for an alcohol or drug problem was considerably higher than the percentage of those who were receiving treatment currently or had received treatment in the past (see exhibit 10). This indicates a significant unmet

Exhibit 10. Self-Reports of Drug and Alcohol Treatment Among Arrestees

	Percentage of arrestees reporting	
Self-reports	Drug misuse	Alcohol misuse
Received treatment in past for drug or alcohol misuse	23	11
Currently receiving treatment for drug or alcohol misuse	13	4
Would like to receive treatment for drug or alcohol misuse	30	15

need for drug and alcohol treatment among arrestees, which is of concern given the high proportion of arrestees, especially IDUs, who saw a link between their drug use and their offending.

XII. Policy Implications and Applications of Data

A number of policy implications flow from this pilot. First, the pilot has shown that the ADAM methodology can be applied in Scotland. It also has shown that a very high proportion (71 percent) of arrestees in at least certain parts of Scotland have used illegal drugs and that nearly one-third (31 percent) may have used opiates. It is not now known whether the close link between drug misuse and crime identified in the two areas of the pilot would be replicated elsewhere in Scotland. The second key policy implication of this research is to underline the importance of extending the ADAM work to other areas within Scotland.

The third key policy implication arising from this limited pilot concerns what one might do to reduce the link between drugs and crime in Scotland. The Scottish Drug Enforcement Agency has recently been created, modeled along similar lines to the U.S. Drug Enforcement Administration. The Scottish Drug Enforcement Agency arose out of a growing recognition of the need to coordinate drug enforcement initiatives at a national level rather than at the level of individual police forces in Scotland (Scotland does not have a national police force, as such, but a collection of regional police forces of which Strathclyde is by far the largest).

The ADAM pilot, however, also has implications in terms of drug misuse treatment and prevention. The results of the pilot illustrate the need to integrate treatment and prevention initiatives better into the criminal justice arena. The ADAM pilot has shown that a high percentage of arrestees perceive a link between their drug misuse and their offending behavior, but few arrestees received help for a drug and/or alcohol problem while in custody. There may be merit in ensuring that drug counselors can access arrestees during their confinement and can provide the necessary support to try to reduce the cycle of offending that so many of the arrestees appear to be locked into as a result of their dependence. Very little is known about the likely success of such an initiative; however, the close link between drug misuse and offending that this pilot has identified in two areas in Scotland underlines the importance of strengthening enforcement, treatment, and prevention initiatives with regard to drug-related offending.

The Future

Following publication of the results of the ADAM pilot, the deputy minister for justice in Scotland announced that there will be a national rollout of the ADAM program in Scotland. Before this takes place, however, the authors are carrying out a scoping study to assess the logistics of initiating a national program involving the interviewing and drug testing of arrestees. This study is investigating sampling methods and what can be done to ensure that a representative sample of arrestees is interviewed, the cost of a nationwide ADAM program, the likely value of combining generic and confirmatory testing of urine samples, and the possible impact on future ADAM work in Scotland of recent legislative changes, including the European Convention on Human Rights.

Acknowledgments

The research described in this chapter was funded by a grant from the Scottish Executive; the views expressed in the chapter are those of the authors and should not be ascribed to the funding body. The interviewers on this study deserve special thanks for the energy and commitment they brought to the project; they were: David Boyd, Anne Broderick, Emily Duizend, Peggy Dwyer, Pablo Jelic, Daniel Martin, Daniel Sumpton, and Shanti Williamson. Dr. Trevor Bennett of the University of Cambridge provided advice to the research team, for which we were very grateful. In undertaking the pilot we owe a debt to Superintendent John McNab (Fife Constabulary) and Inspectors Jo McGallagly and Brian Dunn (Strathclyde Police). The research would not have been possible without the active support of police officers in each of the custody areas or of the arrestees who agreed to be interviewed and to provide urine samples. We are grateful to all of the arrestees who agreed to be interviewed in this research.

Notes

1. Anderson, S., and M. Frischer, "Drug Misuse in Scotland: Findings from the 1993 and 1996 Scottish Crime Surveys," *Crime and Criminal Justice Research Findings*, No. 17, Edinburgh: The Scottish Office, Central Research Unit, 1997.

2. Advisory Council on the Misuse of Drugs, *Tackling Drugs to Build a Better Britain: The Government's Ten-Year Strategy for Tackling Drugs Misuse*, London: Her Majesty's Stationery Office, 1998.

3. Miller, P., and M. Plant, "Drinking, Smoking and Illicit Drug Use among 15 and 16-year-olds in the United Kingdom," *British Medical Journal* 313 (1996): 394–397.

4. McKeganey, N., and J. Norrie, "Pre-Teen Drug Users in Scotland," *Addiction Research* 7 (1999): 493–507.

5. Frischer, M., A. Leyland, R. Cormack, D.J. Goldberg, M. Bloor, S.T. Green, A. Taylor, R. Covell, N. McKeganey, and S. Platt, "Estimating the Population Prevalence of Injection Drug Use and Infection with Human Immunodeficiency Virus Among Injection Drug Users in Glasgow, Scotland," *American Journal of Epidemiology* 138 (3) (1993): 170–181.

6. Hay, G., and N. McKeganey, "Estimating the Prevalence of Drug Misuse in Dundee, Scotland: An Application of Capture-Recapture Methods." *Journal of Epidemiology and Community Health* 50 (4) (1996): 469–472.

7. Frischer, M., D. Goldberg, M. Rahman, and L. Berney, "Mortality and Survival Among a Cohort of Drug Injectors in Glasgow, 1982–1994," *Addiction* 92 (4) (1997): 419–427; Hammersley, R., M.T. Cassidy, and J. Oliver, "Drugs Associated With Drug-Related Deaths in Edinburgh and Glasgow, November 1990 to October 1992," *Addiction* 90 (1995): 959–965; Neale, J., "Experiences of Illicit Drug Overdose: An Ethnographic Study of Emergency Room Attendances," *Contemporary Drug Problems* (Fall 1999): 505–530.

8. Goldberg, D., S. Cameron, and J. McMenamin, "Hepatitis C Virus Antibody Prevalence Among Injecting Drug Users in Glasgow Has Fallen but Remains High," *Communicable Disease and Public Health* 1 (2) (1998): 95–97.

9. Davies, AG., N.J. Dominy, A. Peters, G.E. Bath, S.M. Burns, and A.M. Richardson, "HIV in Injecting Drug Users in Edinburgh—Prevalence and Correlates," *Journal Of Acquired Immune Deficiency Syndromes and Human Retrovirology* 8 (1995): 399–405.

10. Taylor, A., M. Frischer, S. Green, D. Goldberg, N. McKeganey, and L. Gruer, "Low and Stable Prevalence of HIV Among Drug Injectors in Glasgow," *International Journal of STD and AIDS* 5 (1994): 105–107.

11. Hutchinson, S., S.M. Gore, A. Taylor, D.J. Goldberg, and M. Frischer, "Extent and Contributing Factors of Drug Expenditure by Injectors in Glasgow: A Multi-Site Citywide Cross-Sectional Study," *British Journal of Psychiatry* 176 (2000): 166–172.

12. Advisory Council on the Misuse of Drugs, *Tackling Drugs to Build a Better Britain: The Government's Ten-Year Strategy for Tackling Drugs Misuse* (see note 2).

13. Scottish Office, *Tackling Drugs in Scotland: Action in Partnership*, Edinburgh: The Scottish Office, 1999.

14. McKeganey, N., C. Connelly J. Knepil, J. Norrie, and L. Reid, *Interviewing and Drug Testing of Arrestees in Scotland: A Pilot of the Arrestee Drug Abuse Monitoring Methodology*, Edinburgh: Scottish Executive, Central Research Unit, 2000.

15. Bennett, Trevor, *Drugs and Crime: The Results of Research on Drug Testing and Interviewing Arrestees*, London: Home Office, 1998.

16. National Institute of Justice, *Arrestee Drug Abuse Monitoring (ADAM) Program: 1997 Annual Report on Drug Use Among Adult and Juvenile Arrestees*, Research Report. Washington, DC: U.S. Department of Justice, National Institute of Justice, 1998, NCJ 171672.

SOUTH AFRICA

Elna van Niekerk
Charles Parry
Andreas Plüddemann
Antoinette Louw
Gina Weir-Smith
D. Locke

Elna van Niekerk is project manager at the Division of Roads and Transport Technology of the Council for Scientific and Industrial Research in Pretoria, South Africa. **Charles Parry** is director of the Medical Research Council in Cape Town, South Africa. **Andreas Plüddemann** is a scientist at the Medical Research Council in Cape Town, South Africa. **Antoinette Louw** is head of the Crime and Justice Programme at the Institute for Security Studies in Pretoria, South Africa. **Gina Weir-Smith** works at the Human Sciences Research Council, Pretoria, South Africa. **D. Locke** is with the South African Police Service's Crime Information Analysis Centre in Pretoria, South Africa.

I. Overview of Demographic Profile of Nation

The Republic of South Africa occupies the southernmost part of the African continent. The geographic size of South Africa is 1,219,090 square kilometers (757,543 square miles). It has common boundaries with the republics of Namibia, Botswana, and Zimbabwe; the Republic of Mozambique and the Kingdom of Swaziland lie to the northeast. Completely enclosed by South African territory in the southeast is the Kingdom of Lesotho. To the west, south, and east, South Africa borders on the Atlantic and southern Indian Oceans, which gives South Africa a lengthy coastline that is swept by two major ocean currents: the Mozambique Agulhas and the Benguela systems. Isolated, 1,920 kilometers southeast of Cape Town in the Atlantic, lie Prince Edward and Marion Islands, which were annexed by South Africa in 1947.

Exhibit 1 shows general demographic information for South Africa; the country has an estimated population of 43 million people,[1] 54 percent[2] of whom live in urban areas. South Africa's nine provinces have varying degrees of urbanization, from 11 percent in the Northern Province to 97 percent in Gauteng. Economic information is provided in exhibit 2.

Exhibit 1. General Demographic Information for South Africa

Geographic area	1,219,090 km²
Estimated population*,†	43 million
Racial distribution	African, 77%; white, 11%; colored,‡ 9%; Indian, 3%
Religious affiliation	Christian, 68%; Muslim, 2%; Hindu, 1.5%; indigenous beliefs and animist, 28.5%
Gender distribution	52% female, 48% male
Age distribution	≤19 years, 47%; 20–39 years, 35%; 40–69 years, 16%; ≥70 years, 2%
Life expectancy (at birth)	57–64 years
Population growth rate	2.1%§
Birth rate (2000 est.)**	24.56 per 1,000
Total fertility rate (2000 est.)	2.47 children per woman
Death rate	11.89 per 1,000
Literacy rate††	81%
Principal languages	IsiZulu, 22.4%; IsiXhosa, 17.5%; Afrikaans, 15.1%; Sepedi, 9.8%; English, 9.1%‡‡
Urbanization rate	54%
Political subdivisions	9 provinces: Eastern Cape, Free State, Gauteng, KwaZulu-Natal, Mpumalanga, Northwest, Northern Cape, Northern Province, Western Cape

*Updated current population statistics are from *Census 1996 in Brief,* Pretoria: Statistics South Africa, 1998.

†Where not otherwise stated, information is from *Census 1996 in Brief,* Pretoria: Statistics South Africa, 1998.

‡"Colored" now refers to a person of mixed white and African or Malay descent. Historically this group originated in unions between white men and slave women or slave men and Khoikoi or Bushman women. Slaves were from Madagascar, Malaysia, Sri Lanka, and India.

§Source: United Nations, Department of International Economic and Social Affairs, *World Population Prospects,* 1994.

**The 2000 estimates for the birth rate and the total fertility rate are taken from Central Intelligence Agency, *The World Factbook 2000: South Africa,* Washington, DC: Central Intelligence Agency, 2000, retrieved August 20, 2001, from the World Wide Web at www.odci.gov/cia/publications/factbook/geos/sf.html.

††The literacy rate is the percentage of the population aged 15 or older who can read or write. Literacy rates for men and women are roughly equal.

‡‡South Africa has 11 official languages. Other principal languages with fewer speakers include Setswana, Sesotho, Xitsonga, siSwazi, isiNdebele, and Tshivenda.

Crime Rates

A current moratorium on the publication of police statistics prevents analysis of crime statistics for 2000. Some statistics are available for previous years, however. Exhibit 3 presents crime statistics for 1999. Crime levels are measured as a ratio per population and, in the case of car theft and hijacking, the number of licensed vehicles.[3] Between 1998 and 1999, the number of reported violent crimes increased by 9.3 percent. Between 1994 and 1999, this figure increased by 21.6 percent.[4]

Prevalence of Drug Use

Various studies by the Human Sciences Research Council (HSRC) document drug use and its socioeconomic correlates. Findings are based on different household surveys of self-reported use.

Exhibit 2. Economic Information for South Africa

Gross domestic product (GDP) (1999 est.)	$135 billion
GDP growth rate (1999)	1.2%
GDP per capita (1999 est.)	$3,215
GDP composition by sector (1999 est.)	Services, 60%; industry, 35%, agriculture, 5%
Major export partners (1997)	United Kingdom, Italy, Japan, United States, Germany
Major export commodities	gold, diamonds, other metals and minerals, machinery and equipment
Labor force (1999 est.)	15 million economically active
Labor force composition by sector (1997 est.)	Services, 45%; agriculture, 30%; industry, 25%
Poverty rate*	NA
Unemployment rate (1996 est.)†	34%
Human development index (1999)‡	0.702

*As of 1993, the percentage of income and consumption among the lowest 10% of the population in socioeconomic status was 1.4%, and the percentage of income consumption among the highest 10% was 47.3% (Central Intelligence Agency, *The World Factbook 2000: South Africa*, Washington, DC: Central Intelligence Agency, 2000, retrieved August 20, 2001, from the World Wide Web at www.odci.gov/cia/publications/factbook/geos/sf.html).

†The unemployment rate is given as a percentage of the economically active population. In addition, another 26 percent of income earners earn less than R500 (approximately US$73.53 at the 1998 exchange rate) per month.

‡The human development index measures a country's average achievement in three basic dimensions of human development: a long and healthy life, knowledge, and a decent standard of living. The index is a composite index of three variables: life expectancy at birth, educational attainment (adult literacy and the combined primary, secondary, and tertiary enrollment ratio), and per capita GDP (measured as purchasing power parity in U.S. dollars). Countries are grouped into three categories: high human development (≥0.800 on a scale of 0–1.000), medium human development (0.500–0.799), and low human development (<0.500) (United Nations Development Programme, *Human Development Report 2001*, New York: Oxford University Press, 2001, retrieved August 20, 2001, from the World Wide Web at www.undp.org/hdr2001).

Exhibit 3. Urbanization and Incidence of Selected Crimes in South Africa in 1999

	Gauteng Province	Western Cape Province	Northern Cape Province	Northern Province	National average
Percentage of population in urban areas	97	NA	NA	11	54
Highest and lowest incidence of selected crimes per 100,000 population*					
Murder			81	17	55
Robbery w/aggravated circumstances	595		46		226
Residential burglary		1,149		253	603
Car theft	2,081		331		1,541
Car and truck hijacking	507		10		312

*For individual provinces, statistics are given only for the province with the highest and lowest rate for each crime. Crime levels for crimes against persons and residences are measured as a ratio per 100,000 population. For car theft and hijacking, crime levels are measured as a ratio per 100,000 licensed vehicles.

NA = Not available.

Source: ISS, *Comparing Crime in the Provinces*, Nedbank ISS Crime Index, vol. 4, (2000).

Illicit drug[5] use generally manifests as a male phenomenon, but women are overrepresented as users of prescription drugs, such as tranquilizers, amphetamines, and sedatives. Illicit drug use tends to concentrate in urban areas. The ratio of past-year alcohol use to cannabis use seems to be about 10:2. An increase in drug use may have emerged among adults in both the lowest and highest socioeconomic categories. On average, 30.9 percent of African male drinkers and 16.4 percent of African female drinkers may be involved in risky drinking. Cannabis is the most popular illicit drug, especially among offenders and among males in low socioeconomic urban sectors. The use of cocaine, mandrax (or mixtures of mandrax and cannabis), LSD, and heroin also occurs, but to a lesser extent, and it is seemingly more prevalent among subsectors such as offenders.

II. Description of the Criminal Justice System Arrest Process and Detention System

The Criminal Justice System

The criminal justice system, based on Roman Dutch Law and common law, is administered nationally. The Constitution of the Republic of South Africa[6] (Act 108 of 1996) is the supreme law of the country and binds all legislative, executive, and judicial organs of state at all levels of government. The judicial authority of the Republic of South Africa is vested in the courts, which are independent, impartial, and subject only to the Constitution and the law. The administration of justice is a function of the national Government. Fundamental rights (in terms of chapter 2 of the Constitution) bind all legislative and executive bodies of state at all levels of government; they apply to all laws, administrative decisions taken, and acts performed during the period in which the Constitution is in force.

Court structures

The Constitutional Court is the highest court in cases regarding the interpretation, protection, and enforcement of the Constitution. The Supreme Court of Appeal is the highest court in cases dealing with final appeals on all other matters. High courts are divided into 13 divisions. A provincial or local division has jurisdiction in its own area over all persons residing or present in that area, in all causes of action arising and all offenses triable within that area, and over all other matters of which it may take cognizance according to law. These divisions hear matters that are usually of such a serious nature that the magistrate's or regional courts would not be competent to make an appropriate judgment or impose an appropriate penalty. Other courts include special superior courts, regional courts, and magistrate's courts. In terms of statutory law, jurisdiction may be conferred upon a chief (or headman) or his deputy to try and punish an African person who has committed an offense under common law or indigenous law and custom, with the exception of certain serious offenses specified in relevant legislation.

The Attorney-General

The authority to institute and conduct a prosecution in the case of an offense that falls within the jurisdiction of a lower court or the high court is vested in the state. The function of deciding whether or not a prosecution should be instituted is, by law, vested in the attorney-general and the public prosecutors under his or her control. The attorney-general exercises his or her function independently and is accountable only to Parliament. The authority of the attorney-general is delegated to public prosecutors in order to conduct prosecutions in the lower courts.

Juveniles

A child is a person under the age of 18 years. Every child who is in detention shall, in addition to other rights, have the right to be detained and treated in a manner that takes account of his or her age. The South African Law Commission is currently attending to policy and legislation for a new juvenile justice system, under which juveniles can be diverted from the formal court system.

The Arrest Process

The arrest process is based on rules laid down by the Constitution, the Criminal Procedure Act (Act 51 of 1977), other legislation, and a police standing order. As a general rule, the object of an arrest is to secure the attendance of such person at his or her trial. There are, however, exceptions to this rule: arrest for the purpose of further investigation, arrest to verify a name and/or address, arrest to prevent the commission of an offense, arrest to protect a suspect, and arrest to end an offense. The requirements of a lawful arrest are that—

- The arrest must have been properly authorized.
- The individual who effected the arrest must exercise physical control over the person who has been arrested.
- The person who has been arrested must be informed of the reason for the arrest and of the rights that he or she has as an arrested person in terms of the Constitution.
- The person who has been arrested must be brought to the appropriate place as soon as possible.

The information to be furnished to the arrested person is printed in the 11 official languages in the formal police pocket book.

The Detention System

On arrival at the police station, the arrested person must be handed over to the community service center commander or the member in charge of the detention facilities, and the arresting officer should provide information regarding the arrest and complete the Arrest Statement. The fact that the person has been arrested and has arrived at the police station must be recorded in the Occurrence Book. This entry contains the name of the individual who arrested the person; the name of the arrested person; the reason for arrest and, if possible, the case number; and whether the arrested person has any visible injuries, is sick, or is under the influence of intoxicating liquor or any other substance with narcotic effect (in the opinion of the community service center commander). The same information is also entered into the Custody Register.

The community service center commander issues a Notice of Constitutional Rights to every person who is admitted into a police cell. The reference number of the notice, together with other information, is recorded in the Custody Register. A person in custody has the right to choose and consult with a legal practitioner or to have a legal practitioner assigned to him by the state and at state expense. A juvenile (a person below the age of 18) may be detained only as a last resort and should be detained separately from adults.

The arrested person must be brought before a lower court as soon as reasonably possible, but not later than 48 hours after the arrest or, if the 48-hour period expires outside ordinary court hours, the end of the first court day after the expiration of 48 hours.

III. Status of National Drug Policy

In the past, efforts to address substance abuse in South Africa focused largely on control measures that fell under the jurisdiction of the law enforcement and justice sectors. In addition, through welfare agencies such as the South African National Council on Alcoholism and Drug Dependence (SANCA), the state provided resources for the treatment of persons with substance abuse problems. The health and education sectors were minimally involved and scant resources were spent on prevention. Although a number of national plans or strategies to address substance abuse were formulated during the 1980s and early 1990s, the government failed to implement such plans. This resulted in fragmentation of effort and a failure to secure sufficient resources to combat substance abuse.

Under the previous government, South Africa became a signatory to the 1961 U.N. Single Convention on Narcotic Drugs, the 1972 Protocol (which amended the Single Convention), and the 1971 Convention on Psychotropic Substances.[7] In 1992, the South African Drugs and Drug Trafficking Act came into being, making it an offense to supply certain substances to any person knowing or suspecting that the substances will be used for the manufacture of illegal drugs, prohibiting any person from converting property that is known or suspected to be gained from the proceeds of drug trafficking, and making dealing in dangerous and undesirable drugs an offense punishable by up to 25 years' imprisonment. The maximum sentence for possession of drugs is 15 years. There are no prescribed minimum sentences.[8]

Since 1994, Parliament has signed into law several pieces of legislation that have strengthened the legislative basis that would allow the country to sign the 1988 U.N. Convention Against Illicit Traffic in Narcotic Drugs and Psychotropic Substances (1988 U.N. Convention). Two of the most important of these laws are the International Co-operation in Criminal Matters Act of 1996 and the Proceeds of Crime Act of 1996. The former provides formal procedures to be used in obtaining and providing information during a criminal investigation that spans international borders, as well as procedures for the repatriation of criminal proceeds. The latter criminalizes money laundering in general and provides procedures for the restraint and confiscation of criminal proceeds. The Extradition Amendment Act of 1996 further defines extraditable offenses and provides for the designation of states to which extradition may be effected in the absence of formal agreements. The Money Laundering Control Bill of 1997 makes certain bodies and institutions "accountable institutions." Bodies and institutions that receive money on behalf of clients in the normal course of business are required to identify their clients and keep proper records of business transactions with them. Certain transactions that have the potential of being used for money laundering purposes will have to be reported to a central authority[9]; this bill has yet to be passed by Parliament.

In 1995, the National Crime Prevention Strategy (NCPS) was approved by Cabinet. This strategy draws together a range of government departments to develop a holistic approach to fighting crime. Organized crime was listed as one of seven priority areas. The NCPS has been criticized, however, for not adequately confronting the issue of illegal substance abuse and its links to crime. In 1997, the Minister of Safety and Security announced that the Government would introduce strict measures to tighten South Africa's ports of entry in a bid to stem the influx of drugs and arms. He stated that bulk commercial traffic would be cleared through only 19 of the 52 existing border posts and that cargo would be cleared at only 8 of the 36 ports of entry.[10] The Department of Justice has indicated its aim of increasing the number of offenders referred to and entering treatment by way of arrest referral schemes; the court process and post-sentencing provisions and various initiatives are under way to move this initiative forward.

In December 1998, South Africa signed the 1988 U.N. Convention in New York. It was hoped that the Money Laundering Control Bill would be enacted before signing the Convention, but legal opinion was that the legislative basis was sufficient to go ahead without that step having been accomplished. The Prevention of Organised Crime Act of 1998 provides for drastic new powers for police and prosecutors including the forfeiture of criminals' assets on the grounds of "a balance of probabilities" rather than "beyond a reasonable doubt." The focus here is on civil rather than criminal prosecution. The legislation allows the state to confiscate assets gained through illegal means regardless of whether the suspects are convicted. February 2000 saw the first deposit of monies confiscated from alleged drug dealers into the Criminal Assets Recovery Fund. The money is to be used to support law enforcement initiatives. The legislation initially ran into difficulty in the courts, which overturned several forfeitures of property belonging to alleged drug lords by the Scorpions' Asset Forfeiture Unit, resulting in amendments to the legislation.

In terms of regional cooperation, South Africa also signaled the Protocol on Combating Illicit Drug Trafficking in the Southern African Development Community (SADC) Region,[11] which was ratified by Parliament in July 1998. The Protocol provides a policy framework that allows the SADC region to cooperate to ensure that it does not become a producer, consumer, exporter, and distributor of illicit drugs or a conduit for illicit drugs destined for international markets.[12] Although the SADC Regional Drug Control Programme[13] focuses mainly on drug supply reduction, Article 7 of the SADC Protocol gives special attention to demand reduction by requiring member states to "develop, implement and evaluate policies and strategies aimed at establishing a comprehensive and integrated demand reduction programme" that will include the development of community prevention, public and school education, and research activities to address the underlying causes of drug abuse.[14]

Although an accurate determination of the amount of government spending on supply and demand reduction activities is not possible, one can be almost certain that a drop in spending on demand reduction activities relative to supply reduction has occurred during the past few years. Treatment centers have been closed and the plan to handle more substance abuse cases through the primary health care system has not been implemented. Without an increase in the budget for prevention activities, the national and provincial departments of welfare have generally had to reduce their support to nongovernmental agencies (NGOs) involved in prevention activities while initiating few prevention activities themselves. One exception to this outcome is the national "I'm addicted to life" campaign, which was launched in 1995 and is aimed at youths between ages 11 and 20. The campaign uses television, radio, information leaflets, and posters and has been expanded to include video and a teachers' manual.

A major reason for the failure to redirect resources to drug demand reduction activities has been the delay in formulating and implementing a national drug master plan. A committee was established toward the end of 1996 by the national Drug Advisory Board to formulate such a plan; after a lengthy period of consultation and drafting, the National Drug Master Plan (1999–2004) was approved by Cabinet in 1999.[15] Implementation began in 2000 following the appointment of members of the Central Drug Authority.

The vision of the plan is to "build a drug free society together and to make a contribution to [addressing] the global problem of substance abuse"; it focuses on crime, youth, community health and welfare, research and information dissemination, international involvement, and communication. The plan recommends the establishment of—

- A national central drug authority to oversee and monitor its implementation.
- A secretariat to oversee its administration.
- Action committees in 382 magisterial districts to ensure its implementation and the uniform spread of information and policies in every part of the country.
- Provincial substance abuse forums to strengthen member organizations in carrying out their existing functions and to keep substance abuse high on the public and political agenda.

Among the key activities to be implemented within the first year are establishing a secretariat, developing a database of research and a national clearinghouse, developing and implementing a mass media campaign aimed at youths, developing recommendations regarding alcohol advertising, formulating model drug policies for schools, and initiating a radio program aimed at educating the public.

The plan focuses on demand reduction but recognizes the importance of harm reduction without setting out detailed harm reduction approaches. Further research is recommended before reaching any policy decision on the issue of decriminalization of cannabis. With regard to the relationship between drugs and crime, the plan recognizes that drug issues manifest themselves at every level of the criminal justice system but that the precise relationship between drugs and crime is hard to determine. Recognition is given to the fact that there are three main ways in which the two are linked: crimes committed due to the psychopharmacological effects of drugs ingested by perpetrators, crimes committed to feed a drug habit, and crimes committed as a byproduct of involvement in drugs and/or drug trafficking.[16]

IV. Background History to SA–ADAM

In 1996, the Human Sciences Research Council (HSRC), together with the Crime Information Analysis Centre (CIAC) of the South African Police Services, conducted a pilot study in the Hillbrow police station (in Johannesburg), based on the Drug Use Forecasting (DUF) program principles. (DUF was superseded by the ADAM program in 1997.) In 1998, a researcher of the Medical Research Council (MRC) and the Institute for Security Studies (ISS) attended the first annual I–ADAM conference in Miami.

During 1998, the Department of Arts, Culture, Science and Technology extended an invitation to consortiums of research partners to submit substantial proposals for innovative research in various categories, one of which was crime prevention. HSRC joined with CSIR to form a consortium and ISS joined with MRC. A strategic decision was taken to form a consortium of all four organizations—HSRC, CSIR, MRC, and ISS—and to partner with CIAC to develop an integrated proposal based on the ADAM model. This proposal was successful and the project was awarded in December 1998.

The South African ADAM program is based on four waves of data collection: three phases of data collection in three sites, done approximately 6 months apart, and one national data collection phase. The three ADAM sites are:

- Three police stations in the province of Gauteng, within or in close proximity to Johannesburg.
- Four police stations in Cape Town (in the Western Cape Province).
- Two police stations in Durban (KwaZulu-Natal Province).

Cape Town, Durban, and Johannesburg are among South Africa's largest cities. Both Cape Town and Durban are port cities, and Johannesburg, with South Africa's largest airport, is also a common port of entry for drugs. In addition, drug problems are known to be most prevalent in these three cities, and they are already part of an ongoing drug abuse surveillance system, the South African Community Epidemiology Network on Drug Use (SACENDU). The ADAM survey, therefore, would feed into this system. The national survey was conducted in 150 police stations countrywide and based on a stratified random sample.

V. Goals and Objectives

The goals and objectives of the South African ADAM program should be understood within the context of the broad funding principles of the Department of Arts, Culture, Science and Technology. The current undertaking is a 30-month project. The main goal is to establish a South African ADAM program but with some objectives not necessarily directly related to the U.S. ADAM model. The principal aims of this study are listed below:

- Establish a monitoring system based on the U.S. ADAM model to monitor trends in drug use and crime and the crime/drugs nexus.
  —Establish a database to provide tactical and strategic information to combat crime, support intervention strategies, and support policy.

—Identify correlates of substance abuse with other factors (firearms, HIV/AIDS infection).

—Compare different methodologies (national/sentinel, urinanalysis with extrapolation, rural surveillance, alternative drug screening).

- Communicate data to practitioners in an innovative way.
- Explore the relationship between serious drug traffic crimes and other criminal activities.
- Explore how different law enforcement agencies, supported by information from this program, could work cooperatively on drug-related issues.
- Establish a framework of reference and interpretation of arrestee data within the context of other databases.
- Develop a training course for arresting officers regarding arrestees under the influence of drugs.

The study timed data collection as follows:

- The first phase of data collection (sentinel sites or Three Metro Study) took place during August and September 1999, and some results are available.
- The second phase of data collection for the Three Metro Study was scheduled for February and March 2000.
- The third phase of data collection for the Three Metro Study was scheduled for August and September 2000.
- Data collection for the national study was scheduled for February and March 2000.

VI. Getting Started

A. Obtaining Funding

As stated earlier, funding for this project was obtained through the South African Department of Arts, Culture, Science and Technology. Because the project's funding did not come from the most important stakeholders in this project, the challenge of funding a sustained effort remains.

B. Contracting for Data Collection

A request for bids to perform data collection was put out to research/fieldwork/market research companies. Data collection for the national survey was done by a different company than the one that collected data for the Three Metro Study, but the latter was somewhat involved in the areas of training and quality control.

DRA Development was chosen to do the Metro study and A.C. Nielsen to do the national study. The fieldwork companies were selected on the basis of pricing, experience in the field of social research, and capacity.

C. Facility Access

Access to police stations was negotiated through CIAC. Police stations were informed of the survey and instructed to cooperate with field workers. CIAC representatives visited most of the police stations, and the project leaders of the ISS/MRC made personal followup visits. For the most part, the problems experienced dealt with communications (e.g., the duty roster or procedures on arrival at the police station). Results of the August-September survey were shared with the police at the site level to sustain support for future data collection phases.

Access to police stations for the national data collection phase was more complicated from a logistical point of view because high-level personal visits were not always possible. Supervisors of each fieldwork team telephoned the police station and visited at least 1 day before the fieldwork was due to start. CIAC developed protocols to be followed by fieldwork teams in order to ensure security for themselves, members of the police, and people in custody.

D. Knowing the Catchment Area

Brief description of the three sites for the Three Metro Study

Cape Town is a port city on the southwestern coast of South Africa with a population of about 2.7 million. It is the legislative capital of the country and the capital city of the Western Cape Province, which has the second highest per capita income, the highest rate of students passing the national high school matriculation examination, and an unemployment rate of approximately 17 percent. Cape Town has the second busiest harbor in the country and the number of international flights handled by its airport has dramatically increased during the past few years. Approximately 51 percent of the population of Cape Town is colored,[17] with the remainder being African (26 percent), white (22 percent), and Asian (1 percent).[18]

Durban, located on the east coast in the province of KwaZulu-Natal, is a port city with a population of more than 1.8 million persons, of whom 55 percent are African, 24 percent are Asian, 17 percent are white, and 4 percent are colored.[19] Durban hosts the busiest harbor in South Africa and also has an increasingly busy international airport. Unemployment rates are roughly twice those of Cape Town, and rates of poverty are much higher.

Gauteng Province comprises 7.3 million persons; 97 percent live in urban areas. The majority of the residents of Gauteng are Africans (70 percent), followed by whites (23 percent), coloreds (4 percent), and Asians (2 percent). Gauteng is the most economically developed of South Africa's nine provinces; its two main cities are Pretoria (the administrative capital of South Africa) and Johannesburg, a cosmopolitan city that is close to Soweto (a large so-called township). Eight percent of the population of Gauteng Province aged 20 years or older have higher than a grade 12 education. Just over 25 percent of the population is classified as being unemployed, substantially more than in Cape Town. Of employed persons, 16 percent earn less than R500 per month ($73.53).[20]

Police stations were chosen on the basis of their arrest flow (i.e., only stations with larger numbers of arrestees processed weekly were chosen to facilitate the progress of interviewing) and their geographic area (i.e., an attempt was made to cover all population groups by including police stations in various residential areas). Four police stations were chosen in Cape Town, three in Johannesburg, and two in Durban.

National survey

The sampling of police stations was based on the latest available census data. Socioeconomic categories were identified for use in the analysis of arrestee data. The results are representative of the entire nation, including rural areas. Arrest flow data for 1 year have been collected and analyzed.

E. Drug Testing Issues

Panel of drugs tested

The panel of drugs tested included dagga (cannabis), mandrax (methaqualone), cocaine, amphetamines, benzodiazepines, opiates, and LSD. South African drug usage differs from other countries with respect to the high use of mandrax (usually together with cannabis).

The fact that no LSD was found in the sample of 878 arrestees in the Three Metro Study led to discussion about excluding LSD in the national sample. The test for LSD costs double that of the tests for other drugs. A decision was made, however, to include LSD in the national study because the market for LSD may have shifted from urban to rural areas and a national study of this size and budget will possibly never be conducted again and should be as inclusive as possible, within budget constraints.

Laboratory responsible for testing of urine samples

Choice of laboratory in the Three Metro Study was based on the lowest bid (University of Cape Town). The bid for the national survey was awarded to a private laboratory with a national network and infrastructure. Both laboratories used the same drug testing methodologies and cutoff points.

F. Instrument Issues

The survey used two slightly different questionnaires, which have been calibrated to enable comparison.

Cultural applicability

In the Three Metro Study, providing multicultural interview teams solved the problem of cultural diversity. The national study was translated into three languages other than English. The focus was on languages not close to the languages normally spoken, and that may not be as well known to field workers.

Differences between U.S. and South African surveys

The survey used in the Three Metro Study was based largely on items included in the adult ADAM interview, but the South African instrument differs from the U.S. instrument in the following ways:

- Response categories for several items differ to account for different terminology use (e.g., on the item of race).
- Additional items were included in the South African questionnaire to address issues that were felt to be not adequately covered in the ADAM questionnaire (e.g., regarding binge drinking, the type or style of dwelling structure, and knowledge of HIV/AIDS and sexual practices).
- Some items were excluded from the South African instrument, either to allow space for including other items felt to be of greater relevance or because they were thought to be irrelevant in the South African context.
- Some items from the U.S. instrument were included but set out differently in the South African instrument (e.g., on the issues of legal and illegal income and sources of drugs).
- Many items from the U.S. instrument were included but placed in a different position on the South African instrument.
- The South African instrument also included some items from the U.S. Addenda on the main questionnaire (e.g., those relating to firearm use).

VII. Barriers to Developing ADAM

Other than operational restrictions, few obstacles existed to developing the South African ADAM. Because this project ended in March 2001, however, the long-term implementation of the project is the real issue at stake. The main barriers are expected to be:

- **Future funding.** SA–ADAM was conducted with funds from the Department of Arts, Culture, Science, and Technology, which allocates funds to research and other organizations to stimulate new research ideas that may have national value. The Department has formal ownership of the projects it funds only until the funding period of such projects expire.
- **Strategic ownership.** The results from this project can provide useful information to many relevant departments (Safety and Security, Justice, Welfare, and Correctional Services). One of these departments or the Central Drug Authority would have to take ownership of the program to ensure future data collection and analysis.

VIII. National Stakeholders Involved in the Development of I–ADAM

The two main national stakeholders in the development of this project are the Department of Arts, Culture, Science and Technology, which provided the funds for development; and the Department of Safety and Security, which provided the infrastructure of its Crime Information Analysis Centre to ensure access to police stations. Other departments, such as the Departments of Health and Welfare, are also seen as stakeholders to be approached once the results of the project's first phase are known.

IX. Research Methods (Sampling)

Sampling Within Police Stations

The sampling frame included all persons arrested during the time period for which fieldwork was planned (e.g., 1 week). The practical methodology was based on suggestions of the U.S. National Institute of Justice (NIJ): Choose from a calculated proportion of the arrestees already within the holding cells at the start of the shift (using systematic sampling) and supplement this group with arrestees as they are brought in. This was not practical in all cases because the arrest flows were often very low, and interviewers had to take whoever was in the cells and corresponded to the selection criteria of age and time lapsed since arrest.

Sampling of Police Stations

For the national survey, which had to be representative of the country, Statomet (a statistical analysis group affiliated with the University of Pretoria) was requested by the HSRC to design and draw a scientific sample of the catchment area population of police stations. All police areas were classified into 20 homogeneous groups by the HSRC Geographic Information System (GIS) center. A sample of 150 police stations was drawn and a total of 2,000 arrestees interviewed at these stations.

Sample Design

The population consisted of 1,089 points (police stations). The 1998 data of the South African Police Services (SAPS) were used to decide on the number of arrestees to be interviewed at each police station. Police stations were arranged so that, when drawn, a systematic sample would be representative of province, police areas, and the 20 homogeneous groups of police stations.

To make the sample cost effective, a minimum of five arrestees per station were interviewed.

For the urine sample, every third police station was selected. A total of 717 urine samples were taken (budget constraints did not allow 100-percent testing of urine samples).

X. Training Issues

Three Metro Study

All field workers were trained during a 2-day training period in Cape Town. NIJ supported the training by sending experienced personnel to assist. Training covered a study overview, questionnaire guidelines, models for setting up at police stations, urine handling protocols, and shipment of urine samples.

National Survey

Training began on January 24, 2000. Although a different fieldwork company has been contracted for the national study, the company involved in the Three Metro Study was contracted to play a mentoring role in the training process. This decision was made to ensure that the national study would benefit from the learning gathered in the first round of data collection. Both studies produced a training manual.

XI. Data Analysis Results for the Three Metro Study

Demographics

Interviews were done at three sites, which included nine police stations. A total of 970 arrestees were interviewed; of these, 878 (90.5 percent) consented to urine tests (see exhibits 4 and 5 for the number and percentage of arrestees tested at each site). Eighty-three percent of arrestees were male. Most (64 percent) were African, 20 percent were colored, 3 percent were white, and 12 percent were Asian/Indian. More than 70 percent of arrestees were 30 years old or younger (see exhibit 6 for general demographic information about arrestees and exhibits 7 and 8 for information about arrestees by gender and age).

Offense Categories

Violent offenses include murder and attempted murder, assault, weapons, rape (including attempted rape), robbery/theft, and other violent offenses. The category of robbery/theft posed a problem because, although theft is a nonviolent offense, the investigating officer often must decide under which offense to seek prosecution.

Property offenses include shoplifting, theft of motor vehicle, theft from motor vehicle, housebreaking, and other property offenses.

Drug/alcohol-related offenses include drug dealing/possession and alcohol offenses (including driving under the influence of alcohol).

Miscellaneous offenses include illegal immigrant/deportation, fraud, false ID, and others (e.g., obstruction of police, prostitution, gambling, reckless driving).

Prevalence of the main offense categories per site is shown in exhibit 9. For all three metropolitan areas, violent crimes were the most prevalent.

Use of Drugs

Overall, 48.6 percent of arrestees tested positive for any drug. Site-specific figures were 34.8 percent for Gauteng, 52.2 percent for Durban, and 60.1 percent for Cape Town.

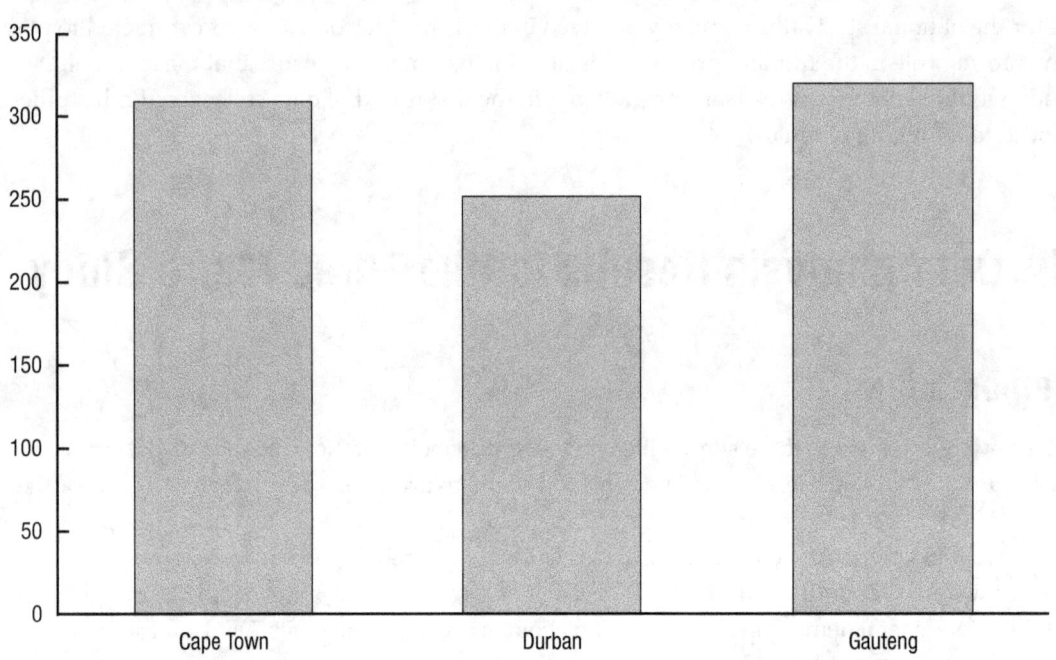

Exhibit 4. Number of Arrestees Tested for Drugs at Each Site

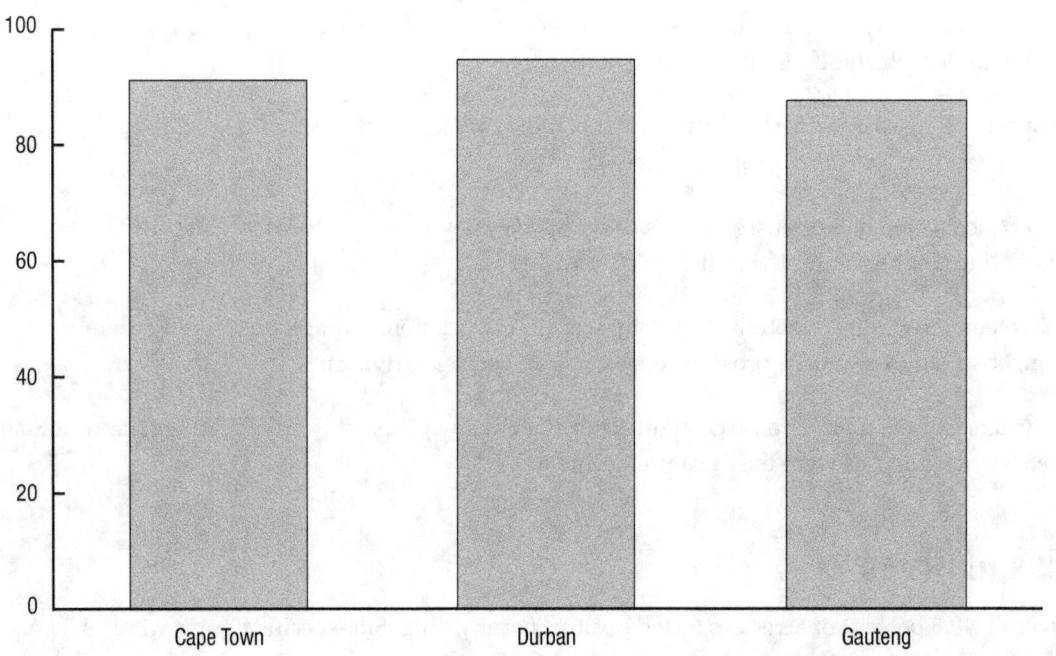

Exhibit 5. Percentage of Arrestees Consenting to Urine Sample

Exhibit 6. Demographics of Arrestees

	Cape Town		Durban		Gauteng		Total sample	
Gender	#	%	#	%	#	%	#	%
Male	291	85.8	206	77.4	306	83.8	803	82.8
Female	48	14.2	60	22.6	59	16.2	167	17.2
Total	339	100.0	266	100.0	365	100.0	970	100.0
Race								
Black	148	45.1	158	59.8	285	84.8	591	63.7
Colored	173	52.7	4	1.5	12	3.6	189	20.4
White	5	1.5	7	2.7	15	4.5	27	2.9
Asian/Indian	1	0.3	95	36.0	11	3.3	107	11.5
Other	1	0.3	0	0.0	13	3.9	14	1.5
Total	328	100.0	264	100.0	336	100.0	928	100.0
Age (years)								
≤20	65	19.3	85	32.0	71	19.5	221	22.9
21–25	99	29.5	66	24.8	106	29.0	271	28.0
26–30	66	19.6	40	15.0	111	30.4	217	22.4
31–35	29	8.6	27	10.2	33	9.0	89	9.2
36–40	40	11.9	18	6.8	19	5.2	77	8.0
>40	37	11.0	29	10.9	24	6.6	90	9.3
Total	336	100.0	266	100.0	365	100.0	967	100.0

Exhibit 7. Demographics of Arrestees by Gender

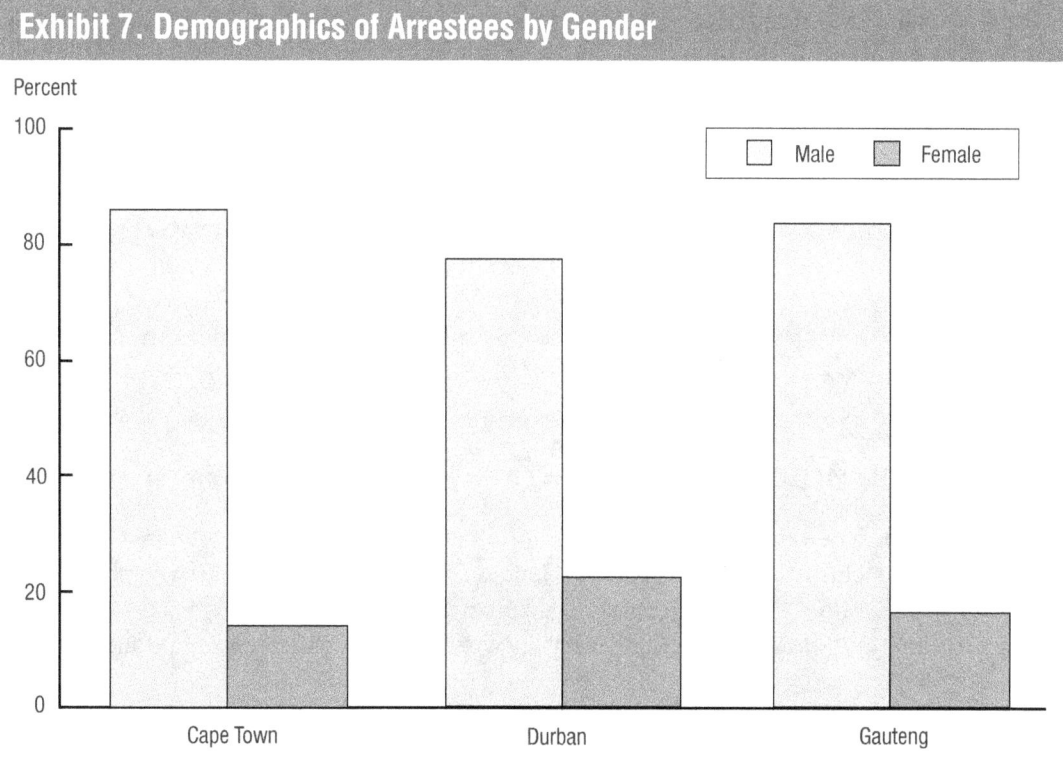

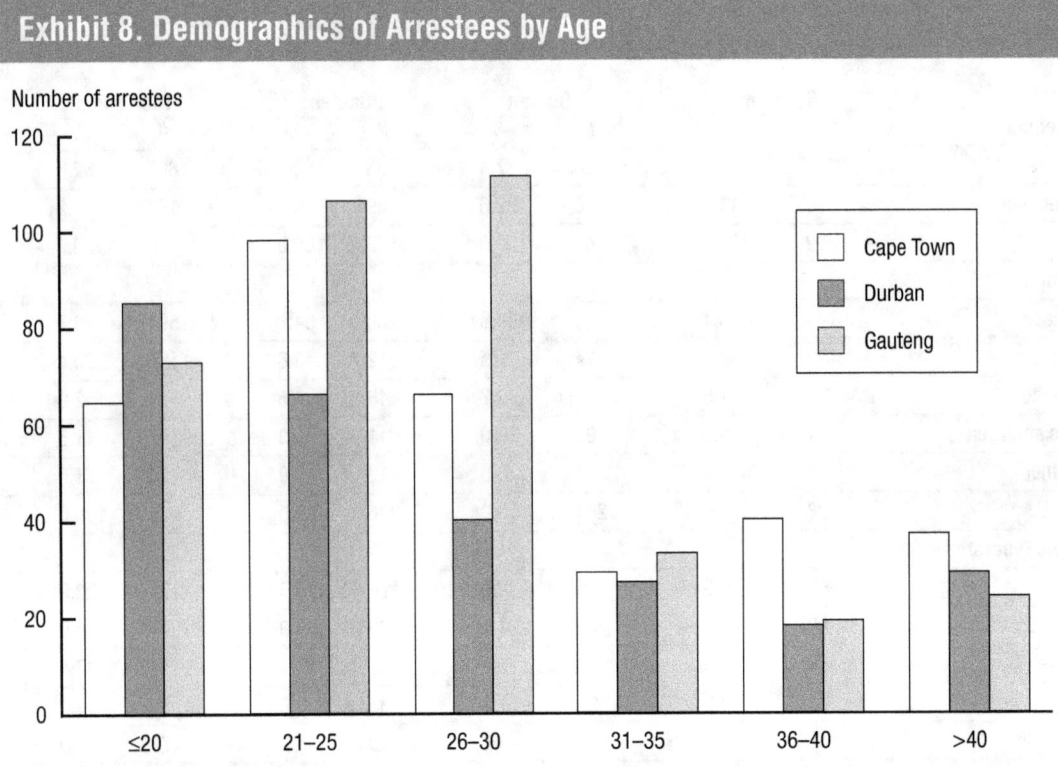

Exhibit 8. Demographics of Arrestees by Age

In Gauteng, especially, a substantial number of arrests were made for immigration offenses (i.e., a person was arrested only for being an illegal immigrant). Because this group of offenders showed relatively low drug use, figures were also calculated with the exclusion of immigration offenses, as follows: 53.7 percent overall, 41.9 percent in Gauteng, 55.5 percent in Durban, and 60.9 percent in Cape Town.

Exhibit 10 shows the prevalence of use of different drugs by offenders, and exhibit 11 shows a breakdown of offenses by drug use and crime type. Seventy-seven percent of housebreaking offenders tested positive for at least one drug, with a high prevalence for dagga (marijuana) of 70.8 percent and for mandrax (methaqualone) of 56.3 percent. Although results were positive for cocaine in 3.7 percent of the tested sample, 12.5 percent of those arrested for housebreaking tested positive for cocaine. Among those arrested for drug/alcohol offenses, 70.9 percent tested positive for at least one drug. Seven percent of those arrested for murder or attempted murder tested positive for cocaine.

Combination Drug Use

In Cape Town, 53.5 percent of the sample tested positive for both dagga and mandrax, followed by 35.9 percent in Durban and 15.3 percent in Gauteng. Drugs that were found in combination were (in numbers rather than percentages) dagga, mandrax, and benzodiazepines (11) and dagga, mandrax, and cocaine (12). Smaller numbers were found for dagga, mandrax, and opiates;

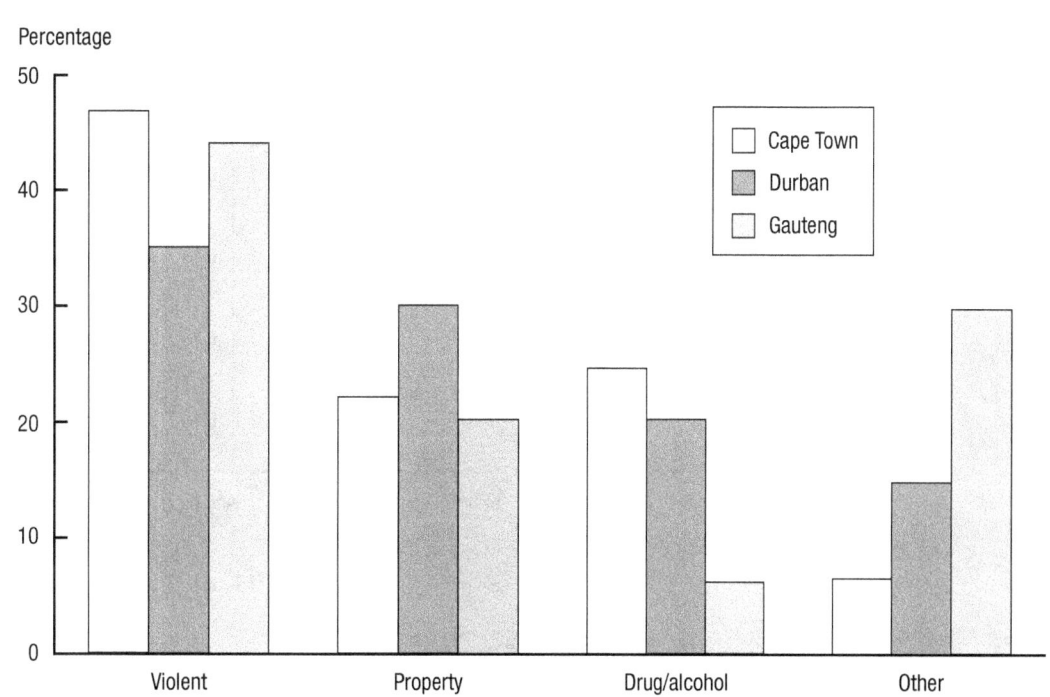

Exhibit 9. Percentage of Arrestees in Offense Categories by Metropolitan Area

mandrax and cocaine; dagga and cocaine; mandrax and benzodiazepines; and a few other combinations. Sixty-three percent of arrestees in Cape Town, 49 percent of arrestees in Durban, and 31 percent of arrestees in Gauteng tested positive for more than one drug.

Comparison Between Phase 1 and Phase 2 Results

The proportion of arrestees testing positive for any drug was fairly consistent over the two study phases. Overall, the proportion of arrestees testing positive for dagga (marijuana) or mandrax (methaqualone) was slightly lower in phase 2 results from 44 percent in phase 1 to 36 percent in phase 2 for dagga, and from 25 percent in phase 1 to 19 percent in phase 2 for mandrax. However, a slightly higher proportion of arrestees (2 percent) tested positive for amphetamines in phase 2, compared with none during phase 1.

Although overall prevalence of arrestees testing positive for any drug and for the various drugs separately was lower in phase 2, some offense categories in phase 2 data showed a higher prevalence of drug-positive arrestees than in phase 1 data. Most notably, the proportion of persons arrested for the motor vehicle theft testing positive for dagga, mandrax, or any drug was considerably higher in phase 2 (74 percent versus 35 percent in phase 1). The proportion of persons arrested for shoplifting testing positive for any drug and for dagga were also higher in phase 2 (37 percent) than in phase 1 (27 percent). The proportion of those arrested for housebreaking testing drug positive remained high (70 percent) in phase 2, although not as high as in phase 1 (74 percent).

Exhibit 10. Percentage of Arrestees Testing Positive for Drugs (Excluding Arrestees for Immigration Offenses)

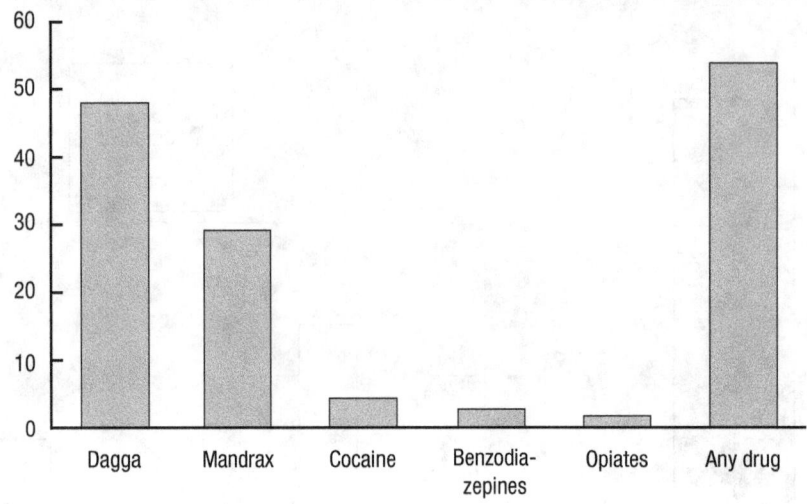

Exhibit 11. Percentage of Arrestees Testing Positive for Selected Drugs by Crime Type

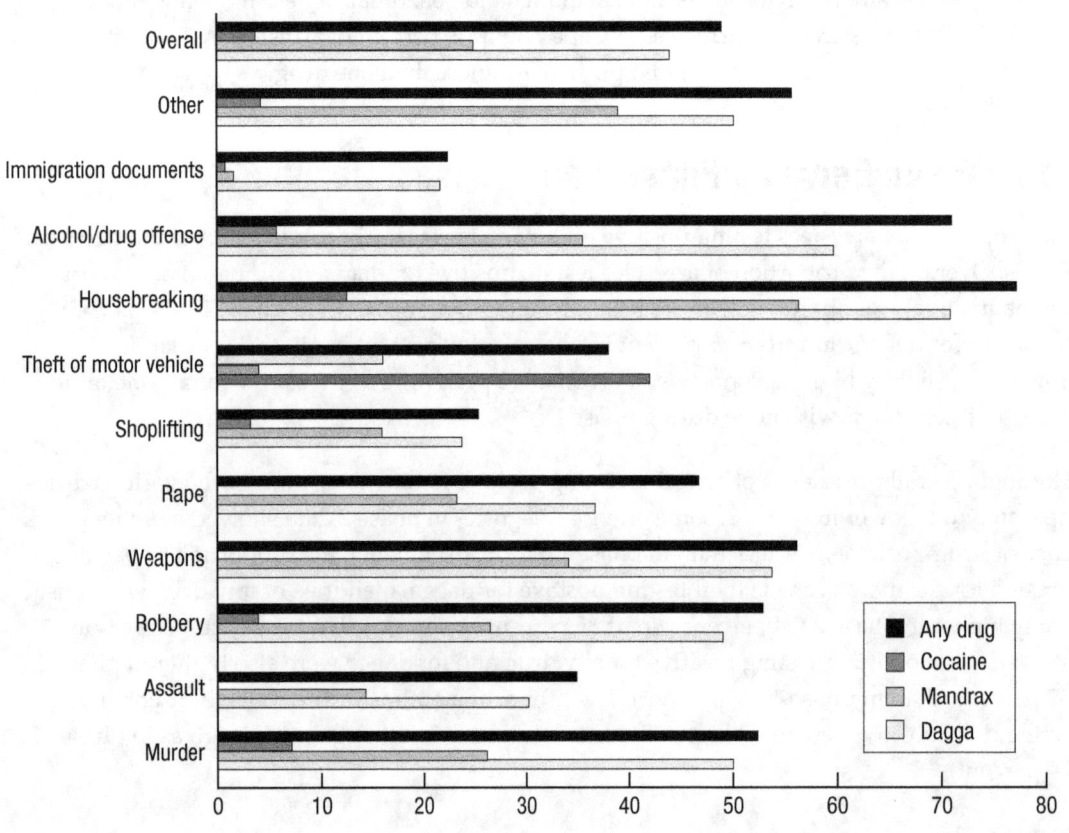

XII. Policy Implications and Applications of Data

Although the data have not been analyzed yet, some thought about their application and policy implications seems appropriate. Specifically, consideration should be given to the following:

- Strategies to reduce drug use and drug-related crime must be area specific.
- Combination drug use seems to be normal for arrestees, especially in Cape Town.
- Crack/cocaine is highly addictive and socially damaging; users should possibly be targeted at police stations.
- Police officers should be sensitive to the fact that the person arrested (especially for housebreaking) is possibly under the influence of dagga and/or mandrax. Police training should include how to recognize the presence of drugs in a suspect and how to handle such a person.
- The crime patterns found are a reflection of police activity.
- The focus on immigration offenses should be reconsidered.
- This project would benefit greatly from the computerization of police arrest data.
- Although alcohol's role in criminal behavior has been better researched and understood than that of drugs, the possibility of alcohol use (alone and in combination with other drugs) should be kept in mind, especially in rural areas, where alcohol is perceived to figure in high violent crime rates.

Notes

1. Updated current population figures by Statistics South Africa, *Census 1996 in Brief,* Pretoria: Statistics South Africa, 1998.

2. Where not otherwise stated, the information is from the 1996 census. Statistics South Africa, *Census 1996 in Brief* (see note 1).

3. Institute for Security Studies, "Comparing Crime in the Provinces: Trends Since 1994," *Nedbank ISS Crime Index* 4 (2000): 7–14.

4. Institute for Security Studies, "A Battle Lost? Violent Crime Trends in 1999," *Nedcor ISS Crime Index* 4 (2000): 1–5.

5. Lee da Rocha-Silva, "The Nature and Extent of Drug Use and Related Problems in South Africa: SARPA National Surveillance Project," Annual Report, South African Researcher Practitioner Association (SARPA), December 1997, available online at http://www.hsrc.ac.za/sarpa/smain.htm.

6. In 1961, when South Africa became a republic, the constitution adopted in 1910 remained in place. In 1983, white voters approved a new constitution that created separate chambers in the legislature for Asians and coloreds. In 1993, a year before the first universal voting in South Africa, a majority rule interim constitution was approved. The current constitution, approved in 1996, was written through a process of consultation and guarantees a broad range of political, social, and economic rights for all the peoples of South Africa.

7. Drug Advisory Board, *National Drug Master Plan*. Pretoria: Department of Welfare, 1999.

8. Omar, D., "The Drug Trade in Southern Africa." Paper presented at a workshop organized by the South African Institute of International Affairs on "The Illegal Drug Trade in Southern Africa," Johannesburg, June 5–6, 1997.

9. Ibid.

10. Kibble, S., *Drugs and Development in South Africa: How Europe Could Help*. London: Catholic Institute for International Relations, 1998.

11. Southern African Development Community, *Protocol on Combating Illicit Drug Trafficking in the Southern African Development Community (SADC) Region*, Gaborone: SADC, 1996.

12. Omar, D., "The Drug Trade in Southern Africa" (see note 8).

13. Southern African Development Community. *SADC Regional Drug Control Programme 1998–2002*, Gaborone: SADC, 1998.

14. Southern African Development Community, *Protocol on Combating Illicit Drug Trafficking in the Southern African Development Community (SADC) Region* (see note 11), 7.

15. Drug Advisory Board, *National Drug Master Plan* (see note 7).

16. Ibid.

17. "Colored" now refers to a person of mixed white and African or Malay descent. Historically, this group originated in unions between white men and slave women or slave men and Khoikoi or Bushman women. Slaves were from Madagascar, Malaysia, Sri Lanka, and India.

18. Statistics South Africa, *Census 1996 in Brief* (see note 1).

19. Ibid.

20. Ibid.

UNITED STATES OF AMERICA

Bruce Taylor

Bruce Taylor is the deputy director of the Arrestee Drug Abuse Monitoring (ADAM) Program in the Office and Research and Evaluation at the National Institute of Justice (NIJ).

I. Overview of Demographic Profile of Nation

The United States of America is a federal republic on the continent of North America. It has an area of 9,629,047 square kilometers (3,717,796 square miles) and is the fourth largest country in the world after Russia, Canada, and China. The United States consists of 48 contiguous States, the noncontiguous States of Alaska and Hawaii, and the District of Columbia. Dependent areas and territories include the Commonwealth of Puerto Rico and the U.S. Virgin Islands in the Caribbean Sea, and the islands of American Samoa and Guam and the Commonwealth of the Northern Mariana Islands in the Pacific Ocean. Political power in the United States is divided between a central governing authority and individual State governments. The Federal Government consists of three branches: executive, legislative, and judicial. Executive power is vested in the President, who conducts the Nation's administrative business with the aid of a Cabinet consisting mainly of the secretaries of the executive departments. The Congress of the United States, the legislative branch, is bicameral, consisting of a Senate and a House of Representatives. The judicial branch comprises the Federal courts. The highest court in the land is the nine-member Supreme Court, which also functions as the court of last resort for the 50 State judiciary systems.

The U.S. population is approximately 281 million, which makes the United States the third most populous country in the world, after China and India (see exhibit 1 for demographic information). Its economy is by far the world's largest, and it ranks sixth in the world in terms of human development (see exhibit 2 for economic information).

Two main sources track victimization and crime in the United States: official police records and research victimization surveys completed by a sample of households. The Federal Bureau of Investigation's (FBI's) Uniform Crime Reports (UCR) program collects information from local law enforcement agencies about crimes reported to police. The UCR crime index includes seven offenses: homicide, forcible rape, robbery, aggravated assault, burglary, larceny/theft, and motor vehicle theft. According to the UCR, the crime index rate fell for the seventh straight year in 1998 and is down almost 14 percent from 1994's rate. Murder was the least frequent violent victimization—there were about 6 murder victims per 100,000 persons in 1998, a rate not seen since the late 1960s.[1]

The National Crime Victimization Survey (NCVS) interviews a nationally representative sample of approximately 49,000 households (about 100,000 persons). In this survey, members of these households who are at least 12 years old are interviewed. The NCVS collects information on crimes suffered by the individuals within those households, whether or not those crimes were reported to law enforcement. Survey data tell how many rapes, assaults, sexual assaults, robberies, thefts, household burglaries, and motor vehicle thefts U.S. residents age 12 or older experience each year. According to the NCVS, U.S. residents age 12 or older experienced approximately 31.3 million crimes in 1998: 73 percent were property crimes, 26 percent were crimes of violence, and 1 percent were personal thefts. Serious violent crime levels decreased from a high of 4.2 million violent victimizations in 1993 to 2.8 million violent victimizations in 1998. Property crime rates also fell, from 552 property victimizations per 1,000 households in 1974 to 217 per 1,000 in 1998.[2]

Self-reported national data on the prevalence of drug use are collected from two main sources: the National Household Survey on Drug Abuse (NHSDA) and the Monitoring the Future (MTF) study. The NHSDA is an annual survey of a nationally representative sample of the civilian, noninstitutionalized population of the United States age 12 and older; 25,500 persons were interviewed in 1998. This survey has been the primary source for estimates of the prevalence and incidence of illicit drug, alcohol, and tobacco use in the United States since the early 1970s. According to the NHSDA, the fraction of today's U.S. population using illicit drugs is well below the peak of the early 1980s. Initiation into the use of illicit drugs rose through the late 1970s and into the early 1980s, then began a sharp decline around 1983. First-time drug use started to rise again in 1992, but in 1997, it was still well below the peak of the early 1980s. Although the number of people who continued using drugs may have risen steadily into the mid-1980s, it almost certainly decreased after 1987. Since 1990, the percentage of the population using drugs has remained essentially the same.[3]

According to the NHSDA, the number of current illicit drug users did not change significantly from 1996 to 1997. In 1997, 6.4 percent (13.9 million persons) of the population age 12 and older reported using an illicit drug during the previous month, down from 14.1 percent (25.4 million persons) in 1979. In 1997, an estimated 0.7 percent (1.5 million persons) of this population group reported using cocaine at least once in the previous month. Overall, the reported use of cocaine, including crack, has not increased significantly since 1991. In 1997, the percentage of Americans age 12 and older reporting the use of marijuana at least once in the previous month was 5.1 percent (11.1 million persons), which was statistically unchanged from 1996.[4]

The other main source of nationally representative drug use data is the MTF study, an ongoing study of the behaviors, attitudes, and values of American secondary school students, college students, and young adults. Each year, approximately 50,000 8th-, 10th-, and 12th-grade students are surveyed (12th graders have been surveyed since 1975, 8th and 10th graders since 1991). In addition, annual followup questionnaires are mailed to a sample of each graduating class for a number of years after their initial participation. At each grade level, the students surveyed are chosen to represent all public and private secondary school students nationwide. Nearly 50,000 students in 422 public and private secondary schools were surveyed in 1998. When this survey began in 1975, 30.7 percent of high school seniors surveyed reported that they had used some type of illicit drug in the previous month. This figure declined gradually to 14.4 percent in 1992 but increased to 25.6 percent in 1998. Previous-month drug use increased from 5.7 percent in 1991 to 12.1 percent in 1998 for 8th graders and from 11.6 percent in 1991 to 21.5 percent in 1998 for 10th graders.[5]

Exhibit 1. General Demographic Information for United States

Geographic area	9,629,047 km² (3,717,796 mi²)
Estimated population (April 1, 2000)	281,421,906
Population by racial group	75.1% white, 12.3% black, 3.6% Asian, 0.9% American Indian and Alaska Native, 0.1% Pacific Islander, 5.5% other, 2.4% multiracial*
Religious affiliation	55% Protestant, 28% Roman Catholic, 2% Jewish, 6% other, 8% none
Gender distribution	50.9% female, 49.1% male
Age distribution	0–14 years, 21.4%; 15–64 years, 66.1%; ≥65 years, 12.4%
Life expectancy (at birth for 2000 cohort)	77.1 years, average; 74.2 years, male; 79.9 years, female
Population growth rate	0.91%
Birth rate	14.2 per 1,000
Total fertility rate	2.075 children per woman
Death rate	8.8 per 1,000
Literacy rate (age 15 and older)	Level 1, 21–23%; level 2, 25–28%; level 3, 31–34%; levels 4 and 5. 18–21%[†]
Language	English
Urbanization rate	89%
Political subdivisions	50 States and the District of Columbia

*As of April, 2000, 12.5% of the U.S. population was of Hispanic origin. The U.S. Census Bureau defines Hispanic as an ethnic, not a racial classification. Persons of Hispanic origin may be of any race (U.S. Census Bureau, Table DP-1, Profile of Demographic Characteristics for the United States: 2000, retrieved August 22, 2001, from the World Wide Web: http://www.census.gov/Press_Release/www/2001/tables/dp_us_2000.pdf).

[†]Kirsch, Irwin S., Ann Jungeblut, Lynn Jenkins, and Andrew Kolstad, *Executive Summary of Adult Literacy in America: A First Look at the Results of the National Literacy Survey,* Washington, DC: U.S. Department of Education, National Center for Education Statistics, 1993. Generally, levels 1 and 2 represent very poor and poor literacy; level 3 represents the minimum level for functioning in an advanced society (high school graduate or college-entry level); and levels 4 and 5 represent command of higher order information processing skills. For a more detailed discussion of the five levels of literacy, see Organisation for Economic Cooperation and Development and Statistics Canada, *Literacy in the Information Age: Final Report of the International Adult Literacy Survey,* Paris: Organisation for Economic Cooperation and Development, and Ottawa: Minister of Industry, Canada, 2000.

Sources: U.S. Census Bureau, Table DP-1, Profile of Demographic Characteristics for the United States: 2000, retrieved August 22, 2001, from the World Wide Web: http://www.census.gov/Press_Release/www/2001/tables/dp_us_2000.pdf; U.S. Census Bureau, *Statistical Abstract of the United States 2000: The National Data Book,* Washington, DC: U.S. Department of Commerce, Economics and Statistics Administration, U.S. Census Bureau, 2001; Kirsch, Irwin S., Ann Jungeblut, Lynn Jenkins, and Andrew Kolstad, *Executive Summary of Adult Literacy in America: A First Look at the Results of the National Literacy Survey,* Washington, DC: U.S. Department of Education, National Center for Education Statistics, 1993.

Exhibit 2. Economic Information for the United States	
Gross domestic product (GDP) (1999 est.)	$9.256 trillion
GDP growth rate (1999 est.)	4.1%
GDP per capita (1999 est.)	$33,833
Median household Income (1999 est.)	$40,816
GDP composition by sector (1999 est.)	Services, 80%; industry, 18%; agriculture 2%
Major export partners (1998)	Canada, 23%; Mexico, 12%; Japan, 8%; United Kingdom, 6%; Germany, 4%; France, 3%; Netherlands, 3%
Major export commodities	Capital goods, automobiles, industrial supplies and raw materials, consumer goods, agricultural products
Labor force (1999 est.)	139.4 million
Labor force composition by sector (1999)	Managerial and professional, 30.3%; technical, sales, and administrative support, 29.2%; services, 13.4%; manufacturing, mining, transportation, and crafts, 24.5%; farming, forestry, and fishing, 2.6%
Poverty rate (1999 est.)	11.8%
Unemployment rate (July 2001 est.)	4.5%
Human development index (1999)*	0.934

*The human development index measures a country's average achievement in three basic dimensions of human development: a long and healthy life, knowledge, and a decent standard of living. The index is a composite index of three variables: life expectancy at birth; educational attainment (adult literacy and the combined primary, secondary, and tertiary enrollment ratio); and per capita GDP (measured as purchasing power parity in U.S. dollars). Countries are grouped into three categories: high human development (≥0.800 on a scale of 0–1.000); medium human development (0.500–0.799); and low human development (<0.500) (United Nations Development Programme, *Human Development Report 2001,* New York: Oxford University Press, 2001, retrieved August 22, 2001, from the World Wide Web at www.undp.org/hdr2001.

Sources: U.S. Census Bureau, *Statistical Abstract of the United States 2000: The National Data Book,* Washington, DC: U.S. Department of Commerce, Economics and Statistics Administration, U.S. Census Bureau, 2001; Dalaker, Joseph, and Bernadette D. Proctor, *Poverty in the United States 1999,* Washington, DC: U.S. Department of Commerce, Economics and Statistics Administration, U.S. Census Bureau, September 2000, Publication P60–210; Central Intelligence Agency, *The World Factbook 2000—United States,* Washington, DC: Central Intelligence Agency, retrieved from the World Wide Web August 22, 2000 at http://www.odci.gov/cia/publications/factbook/geos/us.html.

II. Description of Criminal Justice System, Arrest Process, and Detention System

Origin of the U.S. Criminal Justice System

The U.S. criminal justice system is founded on the principles of English common law and is based on a complex series of laws, procedures, and decisionmaking processes. The system stems from the premise that crimes against individuals are crimes against the state as well and are therefore punishable through state prosecution. The United States does not have a single criminal justice system; systems vary from State to State, county to county, and city to city. All these systems are governed by the broad principles of due process as defined by the U.S. Constitution and as interpreted by U.S. Supreme Court rulings, but legislation and judicial interpretation can vary from State to State.

Types of Offenses

Most offenses committed in the United States are under the jurisdiction of State and local criminal justice systems. Laws are most commonly enforced by cities, counties, and other local jurisdictions; the court systems are most commonly directed by State governments with regional or county jurisdictions; and corrections is primarily the responsibility of county and State governments. Additionally, all but the smallest cities engage in some type of corrections enterprise.[6]

Federal crimes (e.g., kidnaping or transportation of stolen goods across State lines) account for only a small proportion of crimes in the United States. For the most part, State and local authorities investigate and respond to crimes. Of 1,901,773 justice system employees, 8.5 percent (161,058) are Federal and 91.5 percent (1,740,715) are State and local.[7]

Offenses against the state are categorized according to seriousness. The distinction between a felony and a misdemeanor is one of the oldest in the criminal law. In general, a felony is distinguished from a misdemeanor by the punishment that may be exacted for the crime. Conviction on a felony charge usually means that a prison sentence of more than 1 year is authorized. Punishments for felonies range from 1 year in prison to, in some States, a sentence of death. Misdemeanors are usually dealt with more leniently and in many cases do not carry a prison sentence. In general, the maximum prison sentence for a misdemeanor is 1 year in prison plus a fine.

Arrest Through Supervision

An offender's participation in the criminal justice system most often begins when a victim or witness reports a crime to the police. Crimes are also discovered by police officers in the field, through the use of informants, and from investigations and intelligence.

Once it is determined that a crime has been committed, law enforcement personnel will try to identify and apprehend the suspect. In cases in which a suspect is identified, apprehended, and charged with a crime, the police will work in conjunction with the prosecutor (representing the government) to determine whether formal charges will be filed. When charges are not filed, the accused is released.

If formal charges are filed, the suspect must be taken to court, where a judge or magistrate will determine whether the police and the prosecutor have obtained enough information or have "probable cause" to detain the suspect. If there is insufficient information, then the charges will be dropped. If the charges hold, the suspect will obtain or will be assigned counsel to represent him or her throughout the remainder of the process. (A defendant in a criminal case has the right to represent himself or herself, but in the majority of such cases, an attorney will be assigned by the court to help the defendant prepare his or her defense.) The suspect can be released at this point to await his or her next court appearance. Commonly, suspects are released on their own recognizance, to the custody of a third party, or under conditions imposed by the court.

A preliminary hearing may follow the initial court appearance, depending on the jurisdiction. This hearing is held to discover if there is probable cause to believe that the accused committed the known crime. If the judge does not determine probable cause, the case is dismissed. If there is a finding of probable cause, the case will likely go to trial.

The first segment of the adjudication process is the arraignment. The accused is formally charged with a specific criminal violation, advised of his or her rights as a criminal defendant, and asked to enter a plea to the charges against him or her. The defendant can plead guilty, not guilty, or nolo contendere (accepting the penalty without admitting guilt). The judge can either accept or reject the plea. A judge may choose to reject a guilty plea if he or she believes the defendant was coerced or did not understand the consequences of the plea bargain.

If the accused pleads not guilty, the case proceeds to trial. At trial, the prosecution and the defense present their evidence to the judge or a jury of the defendant's peers. Defendants accused of serious crimes have a right to a trial by jury (as do all criminal defendants in Federal court), but even in these cases, the defendant can waive that right in favor of having the judge determine the outcome. The trial will result in a finding that the defendant is guilty or not guilty. If the defendant is found not guilty, he or she is acquitted of all charges and released immediately. If convicted, the defendant is sentenced. Once convicted, a defendant has the right to appeal the conviction. Appeals, however, are often rejected if there is insufficient evidence to show that the offender did not receive a fair trial. In certain cases, such as those in which the death penalty is imposed, the right to appeal is automatic.

The judge most often determines the sentence, but in many States a jury may hand down the sentence at a separate hearing. Sometimes the same jury is used, and in some cases a new panel is convened for sentencing purposes.

Common sentences include fines, restitution, probation, and incarceration. Alternatives to traditional sentencing are often less restrictive than incarceration but are designed to provide more supervision, and are more punitive, than regular probation. In some jurisdictions, the law mandates specific sanctions for certain types of offenses. For example, many States have imposed mandatory sentences for drug offenders.

III. Status of National Drug Policy

The Federal Government plays a significant role in developing drug control policy. On behalf of the President, the Office of National Drug Control Policy (ONDCP) and disseminates broad national policies and strategies designed to affect drug use on a national scale. Additionally, the Federal Government is actively involved in enforcing Federal drug laws through several of its Federal law enforcement agencies, principally the Drug Enforcement Administration, the Federal Bureau of Investigation, and the U.S. Customs Service. The Department of Transportation, through the U.S. Coast Guard, and, to a lesser extent, the Department of Defense are also actively involved in the Nation's counter-drug efforts. Offenders who violate Federal laws are prosecuted in the Federal court system.

In addition to Federal criminal and regulatory law concerning the transportation, sale, possession, and use of drugs, each of the 50 States has enacted its own criminal laws on this subject.

State law enforcement agencies enforce these laws under the direction of the Governor, often through a State attorney general. In many cases, Federal and State law enforcement agencies work together to combine investigative and enforcement resources. In some areas of the country, these joint efforts have been systematized into formalized agreements. Offenders who violate State laws are prosecuted by the respective State court system.

In some instances, both Federal and State drug laws apply to individual violations. In such cases, Federal and State prosecutorial authorities often coordinate with each other as to which level of government will initiate the prosecution of the offender. The caseload of the State and Federal courts and the likely punishment are often determining factors as to who will take the lead.

Federal Responsibilities for Drug Control Policy

The central coordinating agency for U.S. national drug policy is the ONDCP. The Director of ONDCP is a presidential appointee who is confirmed by the U.S. Senate and serves on the President's Drug Policy Council and National Security Council for drug-related issues. Since the late 1980s, ONDCP has been mandated to coordinate Federal efforts to reduce the use of illegal drugs in the United States. The Anti-Drug Abuse Act of 1988, which established ONDCP, requires the Office to develop an annual strategy for reducing illegal drug use and to incorporate goals and measurable objectives for monitoring its progress. In early 1992, underage alcohol use was included among the drugs to be addressed by ONDCP. The Act mandates ONDCP to develop an annual National Drug Control Strategy; a consolidated National Drug Control Budget; and guidelines for cooperation among Federal, State, and local drug control entities. The Act further requires the Director of ONDCP to evaluate, coordinate, and oversee the international and domestic antidrug efforts of executive branch agencies and to ensure that those efforts sustain and complement State and local antidrug activities. The ONDCP Director advises the President regarding changes in the organization, management, budgeting, and personnel of Federal agencies that could affect the Nation's antidrug efforts.[9]

The Violent Crime Control and Law Enforcement Act of 1994 and the Reauthorization Act of 1998 extended ONDCP's mission to include budget and resource powers related to formulating and implementing the President's National Drug Control Strategy, and established new reporting requirements for ONDCP. This new authority gives ONDCP influence over agency budgets to ensure they carry out the priorities, goals, and objectives of the Strategy. ONDCP's reporting requirements under the Crime Control Act include assessing the following areas: prevalence of drug use, reduction in drug availability, and reduced consequences of illicit drug use and availability.[10]

The National Drug Control Strategy is designed to reduce illicit drug use and its consequences, building on five goals first proposed as part of the 1996 National Drug Control Strategy:[11]

- **Goal 1:** Educate and enable America's youth to reject illegal drugs, alcohol, and tobacco.
- **Goal 2:** Increase the safety of America's citizens by substantially reducing drug-related crime and violence.
- **Goal 3:** Reduce the health and social costs to the public of illegal drug use.

- **Goal 4:** Shield America's air, land, and sea frontiers from the drug threat.
- **Goal 5:** Break foreign and domestic drug sources of supply.

The Strategy is intended to prevent and deter first-time use among children and youth, prevent casual users from becoming addicts, remove violent criminals from the streets, dismantle drug networks in the United States and abroad, provide treatment to drug abusers, and reduce the supply of illicit drugs entering the United States. To assess the effectiveness of this strategy, ONDCP has instituted a National Performance Measures of Effectiveness system. This system provides feedback for refining antidrug strategy and allocating resources to antidrug efforts.

IV. Background History of the ADAM Program

The history of the Arrested Drug Abuse Monitoring (ADAM) program can be traced back to the 1980s. In 1984, the National Institute of Justice (NIJ) funded arrested pilot studies in Washington, D.C., and Manhattan to assess the utility of pretrial drug testing and the feasibility and effectiveness of assessing pretrial risk. Together, these two projects involved data collection on 14,000 arrestees.[12] In both studies, urinalysis revealed that more than half of those tested were positive for drugs.[13] This level was far higher than other estimates in 1984 suggested.[14] The results of this pilot work also challenged the validity of self-reported information on drug use and confirmed the usefulness of drug testing for arrestees. More than half of those who tested positive for illicit drug use by urinalysis failed to admit using drugs during a face-to-face research interview.[15] NIJ funded a followup study in Manhattan in 1986 that revealed that cocaine use was rising rapidly; double the percentage of people who tested positive for cocaine in 1984 (42 percent) tested positive in 1986 (83 percent).[16] Had self-reports been relied on exclusively, this increase might never have been detected; the proportion who admitted to ever having used cocaine was about the same for both years.[17]

The high levels of cocaine use detected in these two pilot cities and the steep increase in use over a short period of time (1984–86) raised the question of whether the trend might be an anomaly of these two large East Coast cities. To gauge drug use among arrestees in other urban areas, NIJ, with co-funding from the Bureau of Justice Assistance (BJA), created the Drug Use Forecasting (DUF) program in 1987.

DUF initially collected data in only eight large cities. Adult male arrestees were interviewed in all eight locations and adult female arrestees were interviewed in five locations. In 1988, DUF expanded to include juvenile arrestees in five DUF sites, and a separate interview instrument for juveniles was created. In 1989, the DUF program established 14 new adult male interview sites for a total of 22 sites. The following year (1990), the program expanded to include 21 female sites and 11 juvenile sites. Also in 1990, a special addendum designed to track the spread of "Ice" (methamphetamines) from Hawaii was fielded in the DUF sites. During 1992 and 1993, DUF operations and funding were consolidated within NIJ, and in 1994, DUF moved into NIJ's Office of Research and Evaluation (ORE).

In spring 1995, NIJ proposed the development of the ADAM program. The ADAM program preserves many of DUF's features but includes a call to do the following:

- Expand operations to 75 sites.

- Redesign the data collection and sampling methodology to provide policymakers and practitioners with a rigorous basis from which to assess local arrested drug use and crime patterns.

- Establish an outreach program through which ADAM will monitor the leading and trailing edges of drug problems in targeted populations, such as suburban, rural, or Indian Nation populations.

- Develop local coordinating councils that will generate research addenda, guide the selection of outreach sites, and disseminate each site's findings.

At its core, however, ADAM preserves DUF's simple concept: interviewing and drug-testing arrestees. Between 1995 and 1997, the DUF program carried out a number of projects, including the following:

- Revising the main interview instrument.
- Developing a Spanish version of the interview instrument.
- Piloting TELEDUF (an automated version of the DUF instrument).
- Developing and fielding the Gun Addendum.
- Conducting a special study (in collaboration with ONDCP) on cocaine/crack and heroin drug markets and drug procurement patterns in six DUF sites.
- Studying methamphetamine markets in five western DUF sites.

In 1997, the President included ADAM in the Administration's 1998 budget request to Congress and the formal transformation of DUF into ADAM began. In that year, ADAM began to develop a sampling protocol, review interview instruments, and develop a process for selecting new ADAM sites.

In 1998, 12 new sites were added to the ADAM program for a total of 35 sites (see exhibit 3). During 1998 and 1999, the new ADAM instrument was developed and tested from several perspectives: cognitive testing, experiments on diagnostic validity, and field tests of the interview package.

Other ADAM projects completed during 1998 and 1999 include—

- An analysis of the impact of specimen handling and storage procedures on test results.
- An assessment of opiate testing that confirmed its ability to differentiate heroin use from the use of medications containing opiate compounds.
- Tests for metabolites and markers that indicate crack as well as general cocaine use.

Also during this period, ADAM hosted a series of meetings on publication and communication approaches with local, regional, and national representatives. Finally, this period marked the release of four drug-specific reports that moved away from the community level to assess regional, subgroup, and other issues evident in the trend data.

Exhibit 3. ADAM Sites, 2000

Albany, NY	Des Moines, IA	Miami, FL	Sacramento, CA
Albuquerque, NM	Detroit, MI	Minneapolis, MN	Salt Lake City, UT
Anchorage, AK	Fort Lauderdale, FL	New Orleans, LA	San Antonio, TX
Atlanta, GA	Honolulu, HI	New York, NY	San Diego, CA
Birmingham, AL	Houston, TX	Oklahoma City, OK	San Jose, CA
Chicago, IL	Indianapolis, IN	Omaha, NE	Seattle, WA
Cleveland, OH	Laredo, TX	Philadelphia, PA	Spokane, WA
Dallas, TX	Las Vegas, NV	Phoenix, AZ	Tucson, AZ
Denver, CO	Los Angeles, CA	Portland, OR	Washington, DC

In 2000, NIJ began to incorporate two new elements that significantly strengthen the value of the data collected. First, ADAM introduced a new sampling strategy that provides representative samples at the county level for each site. Second, the new ADAM instrument was implemented. Pending congressional approval, ADAM is expected to begin enrolling 15 new sites in fiscal year (FY) 2002 and 25 new sites in FY 2003 for a total of 75 sites. When fully funded at 75 sites, ADAM will include the largest U.S. cities with populations of 200,000 or more and the largest city in each State where no city has a population greater than 200,000. By collecting drug use data in these sites nationwide, the program will cover more than 40 States, represent more than 20 percent of the U.S. population, and encompass an offender base that is responsible for more than 45 percent of the UCR Part 1 crimes. At full projected size, interviews will be administered to more than 75,000 adult males, 30,000 adult females, and 45,000 juveniles.

V. Goals and Objectives of ADAM

Although the goals of the ADAM program are not formally stated, some of its broad aims can be identified. The ADAM program is an information system on drugs, crime, and related social/public health issues that can be used to meet the following five goals:

1. Provide insights into the link between drugs and crime. By sampling arrestees, the ADAM program can assess the extent to which those involved in criminal activity consume illicit drugs. ADAM data can help enhance understanding of the drug use patterns of actual or potential offenders, the extent to which offenders commit crime to fund drug misuse, and the extent to which these drug-misusing offenders inflate the total number of crimes committed.

2. Monitor the prevalence of hardcore drug use. Research addressing the prevalence of drug use typically does not include those in jail or prison and, therefore, may underestimate levels of drug use in the United States. ADAM is one of several drug indication systems that operate in the U.S. (e.g., National Household Survey on Drug Abuse, Monitoring the Future, Drug Abuse Warning Network). Monitoring drug use among criminal suspects is important not only because of the links between drug use and crime but also because arrestees are likelier than household members and schoolchildren to be heavily drug involved. ADAM contributes to research on the prevalence of drug use by sampling persons who are not sampled by other drug surveys. By focusing on a hardcore population such as arrestees, NIJ can provide estimates of low-base-rate drugs, such as heroin, that are not typically used in the general population. As an ongoing surveillance system, ADAM can monitor changes in drug use over time and compare rates of use across U.S. sites and regions.

3. Assess the dimensions of particular local substance abuse problems at the county level. ADAM is one of the few ongoing drug surveillance systems that provides local estimates of drug use. The vast majority of the crimes committed in the United States fall under the jurisdiction of county or city governments. National, and even regional, drug surveillance statistics tend not to be useful for local policymakers. ADAM responds to the needs of local law enforcement, judicial, and government entities by providing data that is representative of their local areas. The ADAM program has been useful in illustrating local variations in drug use (e.g., methamphetamine use is much more prevalent in many of the western ADAM sites than in East Coast sites).

4. Evaluate, in a low-cost manner, programs and interventions that serve or target criminally active populations. The ADAM program offers a rare opportunity to collect primary data without absorbing the basic operational costs already funded by ADAM. The ADAM data can be used as a source of preintervention data for a before-and-after design. For example, Birmingham, Alabama's Breaking the Cycle (BTC) program is being evaluated through a quasi-experimental design in which the control group is made up of ADAM arrestees selected before the implementation of BTC. Other programs can be evaluated with the ADAM platform. For example, ADAM could be useful for evaluating jail-based programs designed for arrestees, pretrial diversion programs, communitywide initiatives, public awareness campaigns for drugs, and specialized criminal courts. Rigorous experimental evaluations can be integrated into the basic ADAM platform. At the Sacramento, California, ADAM site, an experimental evaluation is being conducted on a treatment program for detained domestic violence arrestees.

5. Provide the necessary data for policy responses appropriate for the arrested population. The ADAM program provides useful drug epidemiological data for many components of Federal, State and local governments that are tackling the problem of drug abuse. Law enforcement, treatment providers, prevention specialists, and other service providers can use ADAM data to assess the impact of their interventions. For example, the congressionally created Methamphetamine Interagency Task Force, co-chaired by the Attorney General and the ONDCP Director, collected information on and proposed strategies to address the growing methamphetamine problem in the United States. ADAM data have been used to support the work of this task force and provide data for the Attorney General to launch a variety of initiatives.

VI. Getting Started

A. Obtaining Funding

The ADAM program is based in the NIJ's Office of Research and Evaluation (ORE). NIJ is based within the U.S. Department of Justice's Office of Justice Programs (OJP). The original funding for the former DUF program was supported by base funds from NIJ's general research budget. In

1998, NIJ received a congressional appropriation to increase ADAM from 23 to 35 sites. Along with a base budget of $2.3 million, ADAM received $4.4 million for a total FY 1998 budget of $6.7 million. This $6.7 million budget remained basically the same in FYs 1999, 2000, and 2001.

ADAM has been designated for expansion to 50 sites in FY 2002 and 75 sites in FY 2003. To obtain the funds needed for expansion, NIJ has assembled a budget proposal that must be approved by officials within OJP, the U.S. Department of Justice, the Office of Management and Budget (OMB), the President, the Senate, and the House of Representatives. Each of these "budget hurdles" requires the careful documentation of past accomplishments and future potential of the ADAM program.

Additional funding for new ADAM sites could come from other Federal sources. In the past, the Centers for Substance Abuse Treatment (CSAT) has funded a number of State data collection programs similar to ADAM. Also, the ADAM program has recently started an ADAM affiliate program. ADAM-affiliated sites operate like other ADAM sites, using all NIJ/ADAM protocols, instruments, and training materials; receiving technical assistance from NIJ/ADAM staff as well as contractor staff; and accessing data from the other ADAM sites. Most of the funding for the ADAM-affiliated sites will not come from NIJ. Funding for these sites may come from State or other Federal sources (e.g., the Federal Byrne grant program). The first ADAM-affiliated site was Albany, New York. Other sites have opened in Charlotte, North Carolina; Kansas City, Missouri; and Tulsa, Oklahoma.

B. Contracting for Data Collection

The ADAM program has four main administrative components: NIJ's internal ADAM program staff (including the ADAM Director), a national data collection contractor (currently Abt Associates Inc.), a laboratory contractor for urinalysis testing (currently PharmChem), and 35 local data collection teams.

NIJ's internal staff

NIJ's internal staff manage the ADAM program, oversee the work of contractors, conduct intramural research, work on ADAM developmental projects (e.g., new survey instruments and using ADAM as an evaluation platform), establish and maintain partnerships with other governmental agencies, and promote the use of ADAM data to a variety of constituencies.

National data collection contractor

The contractor was selected through a competitive solicitation process and is responsible for data collection, consistency, integrity, and management. The contractor collaborates with NIJ on the design and management of ADAM data collection and processing. The current data collection contract is for 4 years and covers responsibilities in the following task areas:

1. Managing all 35 ADAM sites.

- Identify potential sites.
- Assess readiness to collect data.
- Staff all sites through subcontracts with local data collectors (e.g., university data collection teams).
- Train all site personnel.
- Monitor sites for adherence to specified procedures, collection deadlines, data integrity, expenditures, and staff performance.

2. Specific data collection procedures.

- Subcontract with local data collectors to carry out data collection
- Ensure that local data collectors properly identify potential research participants, secure participants' informed consent, conduct interviews with appropriate surveys, and collect urine specimens according to specified procedures.

3. Maintaining and organizing data.

- Edit and enter data into database.
- Process the ADAM urine specimen data.
- Maintain ADAM data files.

4. Analyzing data.

- Prepare the ADAM data findings for semiannual/annual reports, special topic monographs, and proposed specialized research studies/analyses.
- Conduct NIJ-approved studies.

Drug testing laboratory contract

This contract provides for urinalysis testing for all 35 sites, a flexible drug testing program that responds to changing substance abuse trends, and a U.S. Department of Health and Human Services (HHS)-certified laboratory partner that has been tasked with keeping NIJ apprised of changing drug technologies and their relevance to ADAM's policy-related research. The current laboratory contract is for 4 years and covers the following task areas:

- Supply sites with necessary materials (cups, bar coded labels, protective gloves, shipping cartons and materials, mailing labels, etc.).
- Develop and update a specimen handling procedures manual and training plan.
- Analyze urine specimens according to flexible 10-drug panels.

- Maintain routine communications with the data collection contractor and NIJ on data reporting discrepancies.
- Report the results of the drug tests to the data collection contractor and all 35 sites.
- Assist NIJ with special requests to support special research projects (including providing the expertise of a Ph.D. chemist who writes brief technical papers that explain or analyze drug testing).
- Ensure proper quality-control systems to provide valid and reliable urinalysis testing.

Local data collection teams

Each of the 35 ADAM sites have local teams of interviewers and managers that have been hired through subcontracts with the national data collection contractor. Many of these data collection teams come from local universities and research firms. The key personnel that make up the local data collection teams include a site director, a site coordinator, interviewers, and security officers. Their roles are outlined in the ADAM training manual and are briefly reviewed below.

The site director has direct responsibility for all site work, and negotiates and executes the subcontract with the national data collection contractor, establishes and maintains contact with the booking (jail) facilities, hires and maintains adequate staff for project performance, oversees data collection and monitors site adherence to national data collection standards, coordinates and communicates with the national data collection contractor and NIJ on implementing changes to ADAM protocols, oversees proper invoicing and fiscal monitoring of the site budget, and works with the local coordinating council.

The site coordinator has day-to-day operating responsibility for collecting data. This includes scheduling data collection, obtaining supplies, preparing questionnaire materials, training and supervising interviewers, editing data, coordinating the activities of the security officer, packaging and shipping questionnaires and specimens, and filling out all of the necessary log and transmittal sheets on completion of data collection. Although some of these tasks may be delegated to others (e.g., interviewers may help prepare questionnaire materials), the coordinator has operational responsibility for and firsthand knowledge of the data collection process. Some sites require more than one site coordinator.

Interviewers conduct the interviews in a manner that is consistent with the procedures and maintain scientific standards of accuracy and objectivity. Security officers move the arrestees to and from the interviewing area while not discussing the nature and scope of the study. In some sites, a detention facility guard, reserve deputy sheriff, or private security officer is used.

C. Facility Access

Without facility access, there would be no ADAM program. Each U.S. ADAM site voluntarily joined the program through competitive solicitation to NIJ. NIJ funds all elements of data collection, including the cost of officers who provide security for the interviewers. Typically, NIJ, contractor, and local data collection staff have worked together to get the necessary approvals from local government officials for access to specific local facilities. Although the Federal Government

United States of America

does not have the right to mandate participation, NIJ can set the rules for participation in the ADAM program.

To be an ADAM site, each participating jurisdiction and system of facilities needs to be able to—

- Ensure that interviewers have access to all of the facilities necessary to obtain a sample of arrestees from the total arrested population that corresponds with the population of interest to the ADAM program.
- Ensure that site staff have access to information at the time of each interview period that describes the total arrested population so that appropriate arrestees can be selected for interviewing and appropriate statistical adjustments, such as sample weighting, can be accomplished.
- Ensure that the interview staff have access to interview rooms where they can conduct confidential and anonymous interviews and to toilets for collecting urine specimens.
- Make booking data available to ADAM site staff so that an informational cover sheet can be completed prior to the interview.
- Provide arrestees for interviewing within 48 hours of arrest.

The participating facilities are told that they will need to provide this level of access four times per year for approximately 2-week intervals for data collection.

D. Knowing the Catchment Area

To interpret ADAM data, analysts need to understand the population from which the ADAM sample has been drawn. One criticism of DUF was the lack of standardization across the sites and the lack of a consistent catchment area unit. Some sites used city borders and others used county borders to define their samples. The larger county area often drew more nonurban arrestees than if the catchment area had been the city. Without this common unit, comparisons across sites may be difficult or, in some cases, impossible. Another problem with some of the earlier DUF sites was a lack of understanding of the catchment area parameters. In designing the new ADAM sampling scheme, it became apparent that jails that were believed to have arrestees from only one county drew from surrounding counties as well. In these cases, DUF produced city- or county-level estimates that were partially confounded by cases from nonparticipating jurisdictions.

To address this issue, the ADAM program intensively examined the flow of arrestees from the participating counties to ensure that a representative sample of the designated county was drawn. For ADAM to succeed, evaluators need to understand how and when arrestees arrive at the jails and how these factors affect or limit the ability to conduct interviews. Are arrestees held in precinct houses or other facilities prior to booking? When do arrestees make their first court appearance or make bond? The answers to these questions can affect the timing or location of the interview.

A key element of ADAM is that it standardizes data collection at the county level.[18] Therefore, although most ADAM sites are referred to by the name of the largest city in the area, the catchment area of most sites is substantially larger than the named city. Using the county as the

ADAM catchment area offers a number of advantages. First, in many counties, the largest city operates its own jail, and the county sheriff operates a jail for the smaller cities and towns in the county. Often, then, it is relatively simple to expand data collection to the county level. Second, county populations often differ from city populations along such demographic lines as employment, socioeconomic status, and ethnicity. Therefore, county-level data can produce a more mixed population than city-level data can. Finally, counties are easier geopolitical boundaries to track, particularly with respect to the processing of arrestees. The population in city jails tends to change not only because of annexations and population growth but also because cities may start and stop jailing service contracts with smaller surrounding municipalities depending on jail capacity and other factors. County boundaries tend to be more fixed, with little movement of arrestees across county borders. Thus, counties are an easier unit of analysis to monitor over time.

The change from DUF to ADAM has occurred incrementally at many sites. Catchment areas were gradually expanded to the county level during 1998. Therefore, for sites that were not previously operating at the county level, the catchment area may have changed during the 1998 data collection year. In certain complicated sites with multiple jails, sites gradually increased the number of facilities where they collected data as they added interview staff and gained access to additional jails.

E. Drug Testing Issues

As discussed in the Research Methods section of this chapter, drug testing is a key element of the ADAM program because it reveals with a high degree of certainty whether a person has used drugs within 48 hours of testing. During the course of operating the ADAM program, it has been periodically necessary to change laboratory firms that test the urine specimens because Government regulations require rebidding contracts. As a result, the program has had four drug-testing laboratories in the past decade. The most recent change occurred in 1998 when the laboratory services contract was again rebid. In addition, program data collection procedures have changed substantially over time. Through mid-1998, specimens were not shipped for analysis until the end of the 2-week data collection period. There was substantial concern that this delay in shipping could lead to the degradation and decay of urine specimens, resulting in skewed test results. Beginning with the laboratory services contract awarded in 1998, data collection staff were required to send specimens to the laboratory daily or every other day.

NIJ staff undertook a research exercise during the most recent change in laboratory service providers.[19] The goal of the project was to determine the consistency of the urinalysis results when the same urine specimens were analyzed by two independent laboratories (the new and the old laboratory). A total of 931 specimens were collected from four locations and four demographic groups (adult males, adult females, juvenile males, and juvenile females). The samples were analyzed first by the incumbent laboratory and then by a new laboratory using the EMIT™ (enzyme multiplied immunoassay testing) screening assay to test for the presence of the 10 common drugs with the same cutoff levels. The experimental results revealed that the concordance rate was approximately 97.4 percent. The highest concordance rates (100 percent) occurred for methadone, methaqualone, and phencyclidine (PCP).

F. Instrument Issues

The original DUF survey included only basic demographic and drug use questions, because the main focus of DUF was on the urinalysis results. As discussed in the Research Methods section of this chapter, the new ADAM survey includes a wider array of questions than the old DUF survey. There are still a number of inherent limitations in the ADAM data collection environment. First, many things that we would like to know about the arrested population cannot be addressed in this survey. Interview time is limited and certain types of questions cannot be asked in a setting in which extensive development of rapport is not possible. A long and unmanageable survey with many threatening questions will not work in this environment. The process of selecting a limited number of topical areas for the survey included consultation with a variety of policy-makers and drug policy researchers. The ultimate product includes a variety of questions that are important to meet the monitoring needs of policymakers and assist researchers in modeling the dynamics of drug use.

Cultural applicability

The United States is often referred to as the "great melting pot" of different ethnic groups. This mixture of ethnic groups can present a serious challenge to researchers interested in collecting standardized data across 35 different cities. There are often fairly substantial cultural differences among the types of participants that can be found in these 35 sites. A substantial proportion of ADAM participants in the Hawaii site is of Pacific Island descent, and in Los Angeles the largest group of male arrestees is Hispanic. The Pacific Islanders have a very different culture from the Hispanic arrestees.

A variety of assumptions are often embedded within surveys. These assumptions can be based on a set of unfamiliar norms that members of minority groups may not understand. To the extent that members do not understand the ADAM survey, the problem of measurement error is introduced. Therefore, when designing the ADAM survey, careful attention was paid to making sure that it applied to the widest group possible. In some cases, non-English versions of the ADAM survey were developed (currently a Spanish-language version of the ADAM survey is in place). In other cases, extensive pretest consultation was undertaken with members of these minority groups to make sure that the nuances of the ADAM survey would be understood. ADAM staff will continue to monitor and assess the issue of cultural applicability. In some cases, these issues can be addressed through training and related protocols; in other cases, specialized studies will need to be conducted to accurately measure the drug use of these other groups.

Standardization versus local utility

Related to the issue of cultural applicability is maintaining ADAM as a useful local research platform. To the extent that ADAM cannot accurately measure the drug use of local ethnic groups, its local utility is diminished. To allow for cross-site comparisons, ADAM has instituted training procedures and related protocols to ensure that each site collects and measures the same data. The core ADAM survey ensures that sites can be compared across a number of key variables. Through the ADAM addendum program, however, each site is given the opportunity to collect supplementary data outside the core areas. For more than a decade, a number of sites have been collecting data on gun use. Other topical areas for addenda that sites have used include gangs, domestic violence, human immunodeficiency virus (HIV), drug markets, specific drugs, and gambling. As a participant in the International Arrestee Drug Abuse Monitoring (I–ADAM) program, the U.S. ADAM program has agreed to collect a smaller subset of core questions to allow for international comparisons.

VII. Barriers to Developing ADAM

One key issue that NIJ has had to contend with in running its drug monitoring program is revising the former DUF program into ADAM. Building on an established program like DUF has many advantages over creating an entirely new program. Many policymakers and other stakeholders were already familiar with the concept of monitoring the drug use patterns of arrestees. The ADAM program did not need to "sell" people on the importance of the arrested population as an interesting hardcore population worthy of monitoring; DUF had already done that. A number of practices and conventions, however, that had developed over the years with the former DUF program needed to be reformed.

In 1993, the General Accounting Office (GAO) reviewed the major national drug data collection programs (including the former DUF program).[20] GAO's report heavily criticized the sampling methodology of the former DUF program. The problems detailed in the report were major barriers that NIJ had to address. As a response to this report, NIJ reengineered DUF into ADAM.

Although heavily criticized by researchers, the non-probability-based sampling scheme of the DUF program was accepted by many stakeholders at the site level. The rationale for changing to a more representative sampling scheme did not always translate well. The new sampling method sometimes required access to additional jail sites, additional floors or locations within the existing jails, or additional time shifts. These new access issues could be difficult to accommodate at jails where the primary concern was detaining persons suspected of committing crimes. The ADAM program staff and site staff have had to work to convince all of the involved parties of the necessity of the new scheme and why it is worth the extra trouble.

VIII. National Stakeholders Involved in the Development of ADAM

The U.S. ADAM program is based within NIJ, which is the research arm for the U.S. Department of Justice. Within DOJ, NIJ is part of the Office of Justice Programs, which is headed by an Assistant Attorney General.[21] A number of DOJ components are key users of ADAM data, including the Attorney General, the Drug Enforcement Administration, and the Federal Bureau of Investigation. These Justice components have used ADAM data to help allocate resources to drug enforcement and identify drug hot spots.

The ADAM program has developed partnerships with other Federal agencies. NIJ has been working with CSAT to estimate the need for treatment among the arrested population. NIJ has partnered with the Centers for Disease Control and Prevention (CDC) to develop an HIV addendum instrument for use with ADAM arrestees. The addendum focuses on risk behaviors for HIV, sexually transmitted diseases (STDs), hepatitis, and tuberculosis. It includes items measuring unprotected sex and needle sharing, history of HIV testing, and exposure to HIV prevention messages. The HIV addendum was piloted in Miami in December 1999 and three other ADAM sites were designated as data collection sites in 2000. Finally, ADAM works closely with ONDCP on coordinating the interests of NIJ with those of other Federal agencies involved in drug control policy. In 1997, ONDCP helped provide bridge funding for ADAM's expansion from 23 to 35 sites. ONDCP has sponsored a number of joint ADAM-related projects with NIJ, including a study of drug markets[22] and the price elasticity of demand and supply for illicit drugs (grant 99–IJ–CX–K004).

These organizations and others have been important in the development of the ADAM program. In particular, these groups supported budget authority and congressional approval for the expansion of the ADAM sites to 35. Support from the highest levels of these groups and from the Office of Management and Budget (OMB) will continue to be important for the ADAM program as it plans to expand to 75 sites.

IX. Research Methods

The ADAM program has two fundamental components. The first is a questionnaire administered by a trained interviewer to an arrested in a booking facility within 48 hours of arrest. The second is a urine specimen collected from the respondent that is used to corroborate claims about recent drug use. Together, these two techniques provide a powerful research platform.

The Survey

Trained interviewers administer the ADAM questionnaire to arrestees in booking facilities within 48 hours of arrest. Interviews with arrestees are conducted in the jails of 35 participating counties. Information is obtained from respondents with the assurance that it will be kept confidential and will not be used to influence the outcome of their arrest. An informed consent statement is read to all the participants requesting their voluntary participation in an interview and informing them that a urine specimen will be taken. The interviews are conducted by trained personnel over the course of approximately 14 consecutive days.

In 1999, the program began redesigning and testing a new ADAM survey that would focus more specifically on issues of interest to policymakers and practitioners. In FY 2000, the new instrument will allow researchers to estimate the proportion and number of arrestees who use a variety of illegal drugs, estimate the need for treatment and dependence on substances in the arrested population, describe drug markets in an area over time, and establish links to other datasets (such as the National Household Survey on Drug Abuse and the Treatment Episode Data Set) to facilitate development of more comprehensive and integrated national estimates of drug use. Another innovative element of the new ADAM instrument is its use of a "calendar" or "life events" method of collecting data. The calendar method is designed to collect data using a visual cue (a calendar) to better characterize complex behaviors or events and to increase the accuracy of recall. With the new instrument, each face-to-face interview will last approximately 25 minutes (twice as long as the former DUF instrument). A more detailed discussion of the ADAM instrument can be found in the *1998 Annual Report on Drug Use Among Adult and Juvenile Arrestees*.[23]

The Drug Test

Drug testing is an important component of the ADAM program because it reveals, with a high degree of certainty, whether a person has used drugs recently. ADAM is the only national drug research program in the United States that utilizes drug testing. The ADAM program uses the EMIT system to screen for the presence of drugs in urine. The EMIT system provides an inexpensive, yet reliable, drug testing method for screening for a wide variety of drugs. EMIT is highly accurate, with low false-positive rates.[24] Moreover, EMIT has been extensively studied so that cross-reactivity (the sensitivity of the test to drugs that are not actually being tested for) is well documented. The ADAM program uses confirmation testing methods only where the EMIT screening method cannot sufficiently differentiate among drugs that are of interest for policy purposes. For example, policymakers as well as researchers and practitioners have an interest in distinguishing between methamphetamine and amphetamine use. EMIT is specific only to amphetamines as a class of drugs. To determine whether methamphetamine was used, the ADAM program conducts confirmatory testing on specimens that screen positive for amphetamines.

Sampling

Data are collected from male and female arrestees in 35 jurisdictions in the United States. Respondents represent offenders arrested from all crime categories, including property crimes, drug crimes, and violent crimes. The ADAM program introduced a new sampling strategy in 2000 that provides representative samples at the county level and greater ability to make site-to-site comparisons. The DUF program collected data using nonprobability sampling.[25] ADAM's redesigned sampling methodology allows policymakers and planners to generalize ADAM data to all arrest and booking activity that occurs in the catchment area. Use of probability samples provides policymakers with more accurate ADAM estimates and allows them to quantify the uncertainty associated with those estimates. In addition, local policymakers will be able not only to report the proportion of arrestees testing positive for illegal drugs in their counties, but to estimate the number of arrestees who would test positive, which is a powerful tool for resource allocation and planning.

The goal set in establishing sample sizes across sites was to provide an adequate number of cases to each site so that all sites would have a similar level of precision in estimating the proportion of arrestees who test positive for a drug. Cases are sampled with known probability of selection within each 24-hour period (to capture types of cases over the entire day) and on each day of the week (to capture variation in types of cases over different days). Simple sites with a single facility in the catchment area can thus collect a random sample of arrestees, determine the proportion with the characteristic of interest (e.g., dependency for drugs), and, using data on the total number of arrests in the target area for the data collection period, estimate the number of arrestees with that characteristic. The sampling problem is little more complicated in sites with just a few jails. Site staff would sample arrestees from every jail, developing proportionately more cases in large jails than in small jails. In the case of a few jails, the standard error associated with a given sample size would be about the same as in places with a single jail. Sites with many booking facilities, however, have a more complex problem. To capture variation in arrestees across facilities requires that all sites be represented and adequate numbers of cases (representing times and days as discussed earlier) be collected. For sites with many facilities, this requires clustering or grouping them in some logical fashion. This clustering of facilities reduces the precision of the estimates and necessitates collecting a larger sample. In both the simple and complex sites, the overall number of cases is determined in part by the volume of arrest activity. Sites with fewer annual arrests but numerous facilities may collect a larger proportion of their total flow than large, high-volume sites.

The Data Collection Process

Sites collect data on specific populations (adult males, adult females, juveniles) for 2 weeks, four times a year. Data collection is done quarterly for several reasons. First, quarterly data collection helps the program maintain access to jail facilities. Quarterly collection is frequent enough that jail facility staff become familiar with program operations but not so frequent that data collection becomes intrusive. Because continuous data collection is not practical due to the interference it would generate, quarterly data collection ensures that interviewers get frequent opportunities to hone their skills in a live interview environment. Quarterly collection also helps account for potential seasonal variations in arrest and crime patterns that occur in some sites. One byproduct of quarterly data collection is that it generates new information more frequently than many other national data collection programs. Quarterly collection allows policymakers and analysts to view trends as they develop, potentially permitting earlier intervention against problems that may occur.

During data collection periods, a random selection of arrestees is approached and asked to participate in the study. Prospective respondents are read an informed consent statement that notes that no identifying information will be requested and that a urine specimen will be requested at the end of the interview. Bar code strips are placed on both the interview form and the urine specimen bottle to ensure that the interview form is matched to the correct specimen. In some cases where the facility staff permit it, a small incentive, such as a candy bar, is given to respondents who complete the interview. In most sites, more than 80 percent of the individuals approached agree to the interview and, of those, more than 80 percent agree to give the urine specimen. The decision not to collect identifying information is an essential component to achieving such high response rates. With the knowledge that the interview is anonymous and confidential, respondents feel more confident in giving honest answers about sensitive topics.

In addition, the ADAM program ensures that interviewers have no potential conflicts of interests with the respondents. Police officers, whether off duty or on, court workers, and pretrial services staff are not allowed to be ADAM interviewers. Their job-related duties may present potential conflicts of interest with information that may be disclosed during the interview process. Equally important, respondents may be less likely to want to talk to interviewers they see working around the jail in other capacities or who they encounter during pretrial release interviews.

At the conclusion of the interview and the collection of the urine specimen, site staff may ask additional questions of the respondent. It is important that the ADAM interview process be completed prior to asking supplemental questions so that all ADAM arrestees are exposed to the same stimulus and interviewed under the same circumstances. These supplemental questions, called addenda, are usually oriented around a specific topic that policymakers want to address. Examples of addenda administered in the past include: patterns of acquiring and attitudes about firearms;[26] patterns of acquiring and using crack, powder cocaine, and heroin;[27] and patterns of acquiring and using methamphetamines.[28]

X. Training Issues

Training is an important component of the ADAM program. The goal of ADAM training is to convey the practices and procedures required to conduct scientific research surveys to a team of interviewers. Each ADAM site receives training using a standard curriculum and manual that addresses basic and advanced interview skills and site operational procedures. The ADAM training covers the issue of maintaining interviewer safety. Interviewer safety is a team effort between the holding facility personnel and ADAM data collection staff. A key step in the interview process is securing the arrestee's informed consent to participate in the study. The training staff review the concept of informed consent. Each interviewer must understand the principle of informed consent and should be prepared to secure informed consent from the arrestees. Interviewers are required to complete and demonstrate competence in basic interviewing skills and an understanding of project operating procedures. Their continued competence is monitored through quarterly evaluations and site error rates.

ADAM training is conducted by the ADAM national contractor, who works with the local site directors and interviewers to provide one core training per year per site and enhancement training as needed. NIJ oversees the national contractor and sets priorities for the training program.

XI. National Data Analytic Results

For more than a decade, the DUF/ADAM program has published an annual report of the basic drug use prevalence findings (based on urinalysis results) for each of its sites. These reports have shown that the majority of detained arrestees test positive for recent drug use within 48 hours of their arrest. Interestingly, while drug use in the general U.S. population declined dramatically in the mid- to late 1980s, drug use rates seemed to rise among criminal offenders, including arrestees. In Washington, D.C., the first DUF site, the percentage of arrestees testing positive for any drug (mostly cocaine) rose from 55 percent in 1984 to nearly 75 percent in 1989. Throughout the 1990s, however, the rates of detected cocaine use among arrestees stabilized. By 1997, the rates of arrestees who tested positive for any drug were below their highest levels, although the rate of cocaine use remained high.

Aside from playing a key role in tracking hardcore drug use (such as cocaine, heroin, and, more recently, methamphetamine), the ADAM program has provided strong evidence that drug use is subject to powerful cohort effects. For example, cocaine use has declined for the most recent birth cohort of arrestees (born in the 1970s and sometimes referred to as "the blunts generation") but has remained at high levels for older cohorts (born in the 1950s and 1960s). ADAM's trend data show that cocaine-using offenders are getting older. The percentage of 18- to 20-year-old arrestees who tested positive for cocaine in Los Angeles, for example, dropped from 47 percent in 1988 to 24 percent by 1996. There has not, however, been a corresponding decrease in cocaine use by older cohorts (age 30 and over). More recently, ADAM data have helped point out the emerging marijuana epidemic among younger arrestees since the mid-1990s.[29] ADAM data have been very useful for getting drug policymakers to acknowledge this "aging in" and "aging out" process.

The ADAM program has also demonstrated regional variations in the drug problem in the United States. For example, methamphetamine use continues to be seen primarily in Western U.S. cities. In 1998, the methamphetamine urinalysis positive rate in San Diego was more than 30 percent, in Sacramento about 25 percent, in San Jose about 20 percent, in Portland about 20 percent, and in Los Angeles about 10 percent. There is hardly any methamphetamine use among the East Coast ADAM sites, but cocaine is a major problem in these sites. In 1998, the cocaine urinalysis positive rate was more than 50 percent in New York, Atlanta, Philadelphia, and Fort Lauderdale. When the media began suggesting that methamphetamine use would be the drug epidemic of the 1990s, ADAM studies demonstrated no material increase.

The ADAM program has been useful in identifying the prevalence of drug use within a hardcore population of youths. Thirteen ADAM sites also collect data from juveniles (Birmingham, Cleveland, Denver, Indianapolis, Los Angeles, Phoenix, Portland, St. Louis, San Antonio, San Diego, San Jose, Tucson, and Washington, D.C.). Over the years, DUF and ADAM have found that the ADAM sample of youths are generally more likely to use "harder" drugs (e.g., cocaine) than those found in high school samples from the Monitoring the Future study.

The ADAM program has helped many communities understand the dynamics of their local drug and crime problem. For example, an analysis of opiate use among arrestees in New York City focused renewed attention on some of the unique elements of New York's heroin problem. In Los Angeles, specialized ADAM studies have provided a better understanding of the correlations between HIV and STDs and drug use among arrestees. ADAM data were also used to distinguish the differences in use patterns and purchasing practices among crack, powder cocaine, and heroin in Chicago; Manhattan; Portland, Oregon; San Antonio; San Diego; and Washington, D.C. Finally, ADAM data are a key component of the local evaluations of the OJP-funded Breaking the Cycle programs (in Birmingham, Alabama; Jacksonville, Florida; and Tacoma, Washington).

XII. Policy Implications and Applications of Data

As a policy tool, ADAM has been used to assist a variety of practitioner groups. ADAM can help law enforcement better understand changes in local drug markets and how to police these markets. ADAM can be used to help treatment providers plan for the number of requests they may receive for treatment services. The ADAM program has tried to respond to the needs of policymakers by ensuring the information provided is current (ADAM provides more than one data collection point per year), timely (ADAM quarterly reports are delivered to the sites within 30 days after the ADAM Data Center receives all of the necessary site data), and accessible (easy-to-read reports).

ADAM data are invaluable to local, State, and national drug policymakers, courts, law enforcement agencies, treatment providers, and prevention specialists as they seek the most effective and most efficient ways to control illicit drug use and to provide necessary substance abuse treatment. Data collected at ADAM sites inform the decisions of practitioners and policymakers and are used by the Nation's top criminal justice, drug control, and substance abuse researchers as they uncover links between drug abuse, crime, and a variety of other contemporary social issues.

States and localities use ADAM data to better understand the impacts of drug use in individual communities. ADAM data have proved so useful to states and localities that California, Maryland, Missouri, and New York have used other funding sources to expand or replicate the ADAM program. This localized understanding increases the effectiveness of the planning and resource allocation processes. States often use ADAM data in their applications for Federal treatment block and formula grants.

National policymakers use ADAM data to assess larger trends and changes in drug use, drug trafficking, criminal activity, and treatment needs. Many agencies have incorporated ADAM data into their strategic planning processes. ONDCP uses ADAM data to monitor the problem of hardcore drug use and ADAM results are central to its Performance Measurement System. CSAT has used ADAM data in its block grant allocation process and is considering funding programs that correspond to the ADAM protocols to assess local treatment needs. DEA has used ADAM data to guide its efforts in allocating agents to drug hotspots. ADAM data have been linked to the DEA STRIDE (System To Retrieve Information from Drug Evidence) data set to demonstrate that higher illicit drug prices reduce demand for illicit drugs among arrestees. ADAM data are a routine component in the National Institute of Drug Abuse's Community Epidemiology Work Group's assessment of drugs and drug abuse in America's communities. Other Federal users of ADAM data include the Senate Judiciary Committee, the Congressional Budget Office, the National Institutes of Health, the National Drug Intelligence Center, the Bureau of Alcohol, Tobacco and Firearms, and the CDC.

ADAM data are used in a variety of different capacities in a variety of different settings. Between 1995 and 1999, more than 100 articles, projects, and presentations using ADAM data were reported to NIJ. This count does not include the multitude of studies, projects, and reports that cite data from ADAM's annual report or the citations of ADAM data in support of local, State, or Federal program or policy initiatives. The following list illustrates some of the ways ADAM data have affected policymaking, program implementation, resource allocation, and the overall understanding of drugs and crime:

- In Louisiana, State legislation on pretrial drug testing programs was passed in part due to the findings of the ADAM program.
- In Oklahoma City, Oklahoma, the County Community Sentencing Commission used ADAM data to determine the extent of drug use among arrestees before implementing a $4.7 million strategy to take over probation from the State.
- In Indianapolis, Indiana, a special committee convened by the mayor's office that included law enforcement, court officials, and service providers used ADAM data in developing a plan to address prostitution and drug and crime issues.
- In Omaha, Nebraska, the Nebraska Crime Commission and the U.S. Attorney, in response to legislation, commissioned a special research project to assess the growing methamphetamine problem. The findings of this project are being used by the Nebraska State Wide Substance Abuse Treatment Task Force to assess the need for treatment and by other organizations and agencies as they plan responses to drug use issues.
- San Diego, California, continues to use ADAM data to provide direction for its methamphetamine task force.
- In Philadelphia, Pennsylvania, ADAM data guide the deployment of city police officers.
- ADAM data play a key role in direction taken by the Methamphetamine Interagency Task Force, chaired by the Attorney General and the Director of ONDCP.
- ADAM data have been used to determine the effects of policing strategies in reducing drug use and associated criminal activity. In New York City, data were used to assess the effects of "quality of life" enforcement by the New York Police Department.
- CSAT's Substance Abuse Need for Treatment Among Arrestees (SANTA) project used ADAM as its core protocol in early efforts to develop statewide needs assessments.
- Cleveland, Ohio, uses ADAM data in specialized treatment programs for prostitutes.
- Cook County (Chicago), Illinois, and Orleans Parish (New Orleans), Louisiana, use ADAM data in their arrested and pretrial drug testing programs.

Notes

1. Federal Bureau of Investigation, *Crime in the United States 1998: Uniform Crime Reports*, Washington, DC: U.S. Department of Justice, Federal Bureau of Investigation, 1999.

2. Rennison, C.M., *Criminal Victimization 1998: Changes 1997–98 With Trends 1993–98*, Washington, DC: U.S. Department of Justice, Office of Justice Programs, Bureau of Justice Statistics, 1999, NCJ 176353.

3. Substance Abuse and Mental Health Services Administration, *Summary of Findings From the 1998 National Household Survey on Drug Abuse*, Washington, DC: U.S. Department of Health and Human Services, Substance Abuse and Mental Health Services Administration, Office of Applied Studies, 1999.

4. Ibid.

5. Johnston, L.D., P.M. O'Malley, and J.G. Bachman, *Monitoring the Future: National Survey Results on Drug Use, 1975–1999. Volume I: Secondary School Students,* Bethesda, MD: U.S. Department of Health and Human Services, National Institutes of Health, National Institute on Drug Abuse, 2000, NIH publication 00–4802.

6. Clear, T.R., and G.F. Cole, *American Corrections,* 5th ed., Belmont, CA: Wadsworth Publishing Company, 2000.

7. Maguire, K., and A.L. Pastore, eds., *Sourcebook of Criminal Justice Statistics,* 1998, Washington, DC: U.S. Department of Justice, Office of Justice Programs, Bureau of Justice Statistics, 1999, NCJ 176356.

8. In fiscal year 1999, only 33 percent of the Federal drug budget was allocated to prevention and treatment; 67 percent was allocated to supply-side programs. Office of National Drug Control Policy, *1999 National Drug Control Strategy Budget Summary,* Washington, DC: Executive Office of the President, Office of National Drug Control Policy, 1999.

9. Office of National Drug Control Policy, *1999 National Drug Control Strategy,* Washington, DC: Executive Office of the President, Office of National Drug Control Policy, 1999.

10. Ibid.

11. Ibid.

12. Wish, E.D., *Drug Use Forecasting: New York 1984 to 1986,* Research in Action, Washington, DC: U.S. Department of Justice, Office of Justice Programs, National Institute of Justice, NCJ 107272.

13. Toborg, M.A., and M.P. Kirby, *Drug Use and Pretrial Crime in the District of Columbia,* Research in Brief, Washington, DC: U.S. Department of Justice, Office of Justice Programs, National Institute of Justice, 1984, NCJ 94073; Wish, *Drug Use Forecasting* (see note 12).

14. Reardon, J.A., *The Drug Use Forecasting Program: Measuring Drug Use in a "Hidden" Population,* Research Report, Washington, DC: U.S. Department of Justice, Office of Justice Programs, National Institute of Justice, 1993, NCJ 144784.

15. Carver, J., *Drugs and Crime: Controlling Use and Reducing Risk Through Testing,* Research in Action, Washington, DC: U.S. Department of Justice, Office of Justice Programs, National Institute of Justice, 1986, NCJ 102668.

16. Wish, *Drug Use Forecasting* (see note 12).

17. Ibid.

18. Some States, such as Alaska, do not have counties or comparable units of government. In those States, the catchment area is defined by the city or municipal boundaries. The New York City site includes all five boroughs of the city, which are five separate counties.

19. Riley, K.J., N.T. Lu, and B.G. Taylor, "Drug Screening: A Comparison of Urinalysis Results From Two Independent Laboratories," *Journal of Drug Issues* 30 (1) (2000): 171–186.

20. U.S. General Accounting Office, *Drug Use Measurement: Strengths, Limitations, and Recommendations for Improvement*, Washington, DC: U.S. General Accounting Office, 1993, GAO report PEMD-93-18.

21. The National Institute of Justice funds most basic and evaluative research through extramural grants and broad investigator-initiated solicitations or directed solicitations related to such Crime Act initiatives as community policing, violence against women, boot camps/prisons, or drug courts. NIJ also funds intramural research by NIJ staff researchers. NIJ was created by the Omnibus Crime Control and Safe Streets Act of 1968 and was founded in 1968 within the Law Enforcement Administration Agency.

22. Riley, K.J., *Crack, Powder Cocaine, and Heroin: Drug Purchases and Use Patterns in Six U.S. Cities*, Research Report, Washington, DC: U.S. Department of Justice, Office of Justice Programs, National Institute of Justice, and Executive Office of the President, Office of National Drug Control Policy, 1997, NCJ 167265.

23. Arrestee Drug Abuse Monitoring Program, *1998 Annual Report on Drug Use Among Adult and Juvenile Arrestees*, Research Report, Washington, DC: U.S. Department of Justice, Office of Justice Programs, National Institute of Justice, 1999, NCJ 175656.

24. Frederick, D.L., J. Green, and M.W. Fowler, "Comparison of Six Cannabinoid Metabolite Assays," *Journal of Analytical Toxicology* 9 (3) (1985): 116–120; Jones, D.W., D. Adams, P.A. Martel, and R.J. Rousseau, "Drug Population in One Thousand Geographically Distributed Urine Specimens," *Journal of Analytical Toxicology* 9 (3) (1985): 125–130; Willson, N.J., M.J. Kogan, D.J. Pierson, and E. Newman, "Confirmation of EMIT Cannabinoid Assay Results by Bonded Phase Adsorption With Thin Layer Chromatography," *Journal of Clinical Toxicology* 20 (5) (1983): 465–473; Verebey, K., S.J. Mule, J. Alrazi, and J. Lehr, "One Hundred EMIT Positive Cannabinoid Urine Samples Confirmed by BPA/TLC, RIA, and GC/MS," *Journal of Analytical Toxicology* 10 (2) (1986): 79–80; Weaver, M.L., B.K. Gan, E. Allen, L.D. Baugh, F.Y. Liao, R.H. Liu, J.G. Langner, A.S. Walia, and L.F. Cook, "Correlations on Radioimmunoassay, Fluorescence Polarization Immunoassay, and Enzyme Immunoassay of Cannabis Metabolites with Gas Chromatography/ Mass Spectrometry Analysis of 11-Nor-Delta-9-Tetrahydrocannabinol-9-Carboxylic Acid in Urine Specimens," *Forensic Science International* 49 (1) (1991): 43–56; Dextraze, P., W.C. Griffiths, P. Camara, L. Audette, and M. Rosner, "Comparison of Fluorescence Polarization Immunoassay, Enzyme Immunoassay, and Thin-Layer Chromatography for Urine Cannabinoid Screening. Effects of Analyte Adsorption and Vigorous Mixing of Specimen on Detectability," *Annals of Clinical Laboratory Science* 19 (2) (1989): 133–138; Fenton, J., M. Schaffer, N.W. Chen, and E.W. Bermes, Jr., "A Comparison of Enzyme Immunoassay and Gas Chromatography/Mass Spectrometry in Forensic Toxicology," *Journal of Forensic Science* 25 (2) (1980): 314–319.

25. Evidence suggests that DUF program interviewee samples were fairly representative of their populations. A 1993 study that investigated the sampling method used by the DUF program concluded that "the current DUF procedures appear to select a sample of interviewees that is highly representative of the arrestees who are detained in the particular booking centers where the DUF program operates" (Chaiken, J.M., and M.R. Chaiken, "Understanding the Drug Use Forecasting (DUF) Sample of Adult Arrestees," unpublished report prepared for the U.S. Department of Justice, National Institute of Justice, Washington, DC, 1993, p. 45). The authors of the study note,

however, that the samples are not wholly representative of all UCR arrestees because the DUF selection procedures rule out minor offenses.

26. Decker, S.H., S. Pennell, and A. Caldwell, *Illegal Firearms: Access and Use by Arrestees,* Washington, DC: U.S. Department of Justice, Office of Justice Programs, National Institute of Justice, 1997, NCJ 163496.

27. Riley, *Crack, Powder Cocaine, and Heroin* (see note 22).

28. Pennell, S., J. Ellett, C. Rienick, and J. Grimes, *Meth Matters: Report on Methamphetamine Users in Five Western Cities,* Research Report, Washington, DC: U.S. Department of Justice, Office of Justice Programs, National Institute of Justice, 1999, NCJ 176331.

29. Golub, A., and B. Johnson, "The Marijuana Upsurge," unpublished final report submitted to the U.S. Department of Justice, Office of Justice Programs, National Institute of Justice, Washington, DC, 2000.

Concluding Thoughts

Henry H. Brownstein

Henry H. Brownstein is the director of the Drugs and Crime Research Division of the National Institute of Justice.

On a political level, the problem of drug using and trafficking is an international problem. In terms of social welfare and public health and safety, the problems associated with drug using and trafficking are local problems. That apparent paradox is precisely what makes the International Arrestee Drug Abuse Monitoring (I–ADAM) program so interesting and valuable. In participating nations, I–ADAM provides data about local problems with drugs and crime and offers an international context through which to understand and explain those local problems as they relate to the global drug problem.

For 3 days in September 2000, representatives of 11 different countries from 6 different continents met in Washington, D.C., at an I–ADAM conference sponsored by the National Institute of Justice of the U.S. Department of Justice. Present were researchers and policymakers actively working on ADAM-type programs in their own nations and researchers and policymakers from nations that were considering the development and implementation of such a program.

For 3 days, those present at the third annual I–ADAM meeting in Washington talked about the status of the project in their own countries, data availability and the potential for sharing data, the data collection instrument and how it could be improved, the different laboratory methods for drug testing, and the future of the I–ADAM program. In turn, participants spoke about what they were learning about drug use among arrestees in their own countries. To the extent possible, findings from data analysis in each country were presented and shared. What was remarkable was how similar and yet how different were the problems and concerns in different nations.

Sharing Information

The different drug problems of different nations were plainly delimited in presentations at the I–ADAM meeting in September 2000. Everyone at the meeting agreed that data from an ADAM-type program provides useful information that can inform local policy and practice. It also became clear, however, that given different systems of criminal justice in different nations, each nation faces its own limitations and difficulties in establishing and operating an ADAM-type program.

Representatives from nations with existing ADAM-like programs and those from nations considering or planning such programs indicated similar levels and types of interest in the program. Vicknasingam Balasingam in Malaysia, where the program is "in its infancy," noted that his minister of justice and the local police chief supported a program that would substantiate claims that crimes are committed by drug addicts and provide information to local health and enforcement agencies. In Taiwan, where the program is still in the planning stage, Shu-Yu Lyu reported that it has support from the Ministry of Justice, the National Police Administration, the Department of Health, and the Ministry of Foreign Affairs. Daleen Lock reported that the South Africa ADAM (SA–ADAM) program, which was piloted in 1996, has been found useful by the police for narrowing the focus of their efforts to deal with the problems of drugs and crime.

According to the I–ADAM presentations, of all illicit substances used by people in all nations of the world, marijuana in one form or another is the drug that is most commonly used by arrestees throughout the world. For example, Toni Makkai reported that, in Australia's Drug Use Monitoring in Australia (DUMA) sites, from 50 to 66 percent of all arrestees tested positive for cannabis, 40 percent tested positive for opiates in Sydney sites, and 20 percent tested positive for opiates in East Perth and Southport. Luis Caris reported that in Chile, alcohol was the primary drug used by arrestees, but of all illicit drugs, marijuana was used most often (by 30 percent of arrestees).

Although some drug problems are commonly shared by many nations, others are peculiar to particular countries. Trevor Bennett reported that in England, across the sites in the NEW–ADAM program (New England and Wales ADAM), from 14 to 50 percent of arrestees tested positive for opiates. In South Africa, Andreas Plüddemann reported that, similar to other nations, from 36 to 44 percent of arrestees tested positive for cannabis. However, from 19 percent to 25 percent of arrestees in South Africa tested positive for mandrax, a blend of methaqualone and antihistamine that is rarely, if ever, found being used by arrestees in other nations. Neil McKeganey reported that, in Scotland, drug injection is a particularly serious problem. Apinum Aramrattana reported that the main problem in Thailand is opiates.

In addition to the similarities and differences in findings, those present also reported differences and similarities in terms of the methodologies nations use and the problems they have in implementing I–ADAM. For example, in all countries people commit crimes and are arrested and booked without regard to the time, yet in most countries, including the United States, it is not possible to conduct interviews 24 hours every day. In England, interviews are collected for 24 hours, but it is not unknown for a respondent to doze off during an interview (this may be due to the high proportion who test positive for opiates). Gina Weir-Smith reported that SA–ADAM researchers in South Africa have been combining databases using artificial intelligence to try to develop strategies for extrapolation of findings to the whole nation. Natalie Lu of the U.S. ADAM program reported on the results of an experiment that showed that different laboratory tests used by different nations for testing urine for drugs produce similar results.

Writing chapters for this compendium offered researchers working on ADAM-like programs around the world an opportunity to focus on their own programs in the context of their national system of criminal justice. Consequently, many of the same researchers who spoke at the I–ADAM conference were able, for this compendium, to provide greater detail and more insight into the operation and output of their ADAM effort. The result is that a number of patterns or themes about both drug problems and the study of drug problems in different nations can begin to be seen.

Each chapter in this compendium outlines the character and principles of the criminal justice system in that nation, the processes and problems of establishing an ADAM-like program, and to the extent they are available, findings and policy implications of the effort to date. For example, in Australia there is no central facility for detainees in Sydney, one laboratory is used to conduct EMIT™ (enzyme multiplied immunoassay testing) testing of urine samples, and arrestees who test positive for substance use most commonly test positive for cannabis and opiates. In Chile, where nurses from the Ministry of Health conduct the interviews, arrestees most often test positive for alcohol, marijuana, and coca paste (more often than they do for powder cocaine). In England and Wales, tests are most often positive for cannabis, alcohol, opiates, and, to a lesser extent, cocaine, and about 23 percent of arrestees admit to having ever injected at least one substance. Malaysia has yet to implement its planned ADAM program, but it will do so in a nation with laws that are particularly tough with regard to drug dealers. In the Netherlands, where drug policy is focused on health issues, arrestees most often have been found to be using cannabis or cocaine at the time of their arrest. In Scotland, where the local police have been actively involved in the program from the beginning, opiates seem to be a problem. In South Africa, where the criminal justice system is administered nationally, the problem drugs among arrestees appear to be marijuana and mandrax. In the United States, where ADAM programs are operating in numerous sites around the country, marijuana appears to be used by large proportions of arrestees in most all sites and there appears to be regional variation in the use of other drugs.

Generally, the chapters in this compendium show that researchers in those nations where I–ADAM programs have been implemented or planned typically have found high levels of interest in the program among policymakers and practitioners in the areas of criminal justice and public health. At the same time, most are experiencing problems identifying and securing adequate funding to support an ongoing program. All must deal with collecting data in an environment that is inherently hectic, and all must be able to work with local law enforcement and correctional officials to ensure that the local ADAM program operates in a way that benefits the community without disrupting the ongoing operation of the local jail. Finally, all must be able to produce findings that can be translated into conclusions that are interesting and valuable to the local community.

Moving Forward

I–ADAM is a network of researchers from various nations working together to address problems in each of their countries that are unique in some ways yet common in others. The drug problem is global, yet the problems associated with drugs in individual nations are distinctive to the political, social, and cultural life of that nation. Through statistical modeling, I–ADAM can produce some interesting comparisons about drug use among arrestees in different countries. I–ADAM allows researchers to inform policymakers and practitioners in their own countries about the problems of drugs and, in turn, to be informed by an international community of researchers who are committed to addressing those same problems in their own countries.

On the last day of the 2000 I–ADAM meeting in Washington, D.C., everyone at the table came to recognize the same potential for the program. The focus of the meeting turned to finding ways that the researchers involved in these efforts could begin to communicate with each other more routinely and more broadly. The result was the formation of a number of what might be called "international cybercommittees." Those present all agreed at some level to participate through the Internet as a member of such a committee in one or more areas. On an organizational level, one group would consider strategic planning and coordination, a second methods and technical support, and a third publications and dissemination. A mission was identified for each group.

It was decided that the mission of the strategic planning and coordination group was to ensure the future development and policy relevance of the work of I–ADAM, and the connectedness of the mutual efforts to national goals. To these ends, in the months following the meeting, several researchers from a variety of involved nations began to consider common themes for cooperative study (e.g., violence and drugs) and different ways to communicate. These themes and plans will be integrated into the planning for the next I–ADAM meeting.

The methods and technology group established a multifaceted mission that included helping I–ADAM participating nations and others planning to participate through advice, research design support, encouragement, facilitation of implementation, promotion of common methods, and mutual planning. In effect, there were two ideas. One was to find within the context of each national I–ADAM effort something that several, if not all, nations could do in the same way that would allow an international comparison. For example, following the meeting, one member of the group proposed expanding the current core instrument so comparisons could be done. The second idea was to ensure that participating nations share what they have learned with other nations planning to start an I–ADAM program. This is being done through meetings and through e-mail communications.

Perhaps the most work is being accomplished in the area of publication and dissemination. The mission of the publications and dissemination group at the meeting was defined in terms of planning, encouragement, and promotion of information from I–ADAM that addresses the problems of drugs and crime. This compendium is certainly an important step in disseminating the information that is being learned from I–ADAM. There are others as well. A number of I–ADAM researchers have published reports and articles from findings in their own countries that are shared with researchers in other nations. In addition, since the meeting in Washington, a number of researchers have been regularly communicating by e-mail to plan joint publications. Trevor Bennett and Bruce Taylor have been working together to prepare a multiyear publication plan.

Concluding Thoughts

In reality, it is not easy for people from different countries with obligations in their own nations to not only communicate but also plan and cooperate to do things beyond the scope of their normal responsibilities. In that sense, the progress of I–ADAM planning may be moving more slowly than one would like. But in that same sense, it is remarkable that anything is getting done. The fact that plans and publications exist is a tribute to the commitment of the people who are involved in I–ADAM. Ultimately, I–ADAM is a fragile multinational network of people who share a common interest in the problem and problems of drugs and are trying to work together to do something about it.

Appendix A

Model I–ADAM
Survey Instrument

[sponsoring agency] 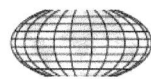 **ADULT I–ADAM INTERVIEW** **English Language Version** DRAFT	NOTICE: Information contained on this form that would permit identification of any individual or establishment has been collected with a guarantee that it will be held in strict confidence, will be used only for purposes stated for this study, and will not be disclosed or released to others without the consent of the individual or the establishment in accordance with *[cite relevant legislation]* Collection of information is estimated to average 15 minutes per interview. Send comments regarding this estimate or any other aspect of this collection of information, including suggestions to *[cite relevant contact]*		
	INTERVIEW DATE ___/___ ___/___ ___ month / date / year	SITE ID # ___ ___ ___	ANONYMOUS STUDY ID# *[PLACE BARCODE STICKER HERE]*

INFORMATION FROM RECORDS
(COMPLETE BEFORE APPROACHING ARRESTEE)

Year of Birth: ___ ___

Sex: 1 - Male 2 - Female

Ethnicity Information: *[use locally appropriate categories]*

1 - 2 - 3 - 4 -

5 - 6 - Other: Specify _____

Precinct/location of arrest: _____
Location of arrest postal code: ___ ___ ___ ___
If not postal, specify nearest intersection to arrestee's residence: _____ and _____

Arrestee's residence postal code: ___ ___ ___ ___
If not postal, specify nearest intersection to arrestee's residence: _____ and _____

Was the person charged with a warrant <u>only</u>? 0 - NO 1 - YES
Was the person charged with a probation/parole/release on own recognizance violation? 0 - NO 1 - YES
Booking Facility code: 1 2 3 4 5 6 7 8 9 0

(WRITE IN CHARGE, WITH NO ABBREVIATIONS)	Charge Code (from list below)	Penal Law Code	Misd	Felony	Status
Most serious charge: _____	_____	_____	M	F	S
Second most serious charge: _____	_____	_____	M	F	S
Third most serious charge: _____	_____	_____	M	F	S

VIOLENT OFFENSES

1 01 Assault
1 02 Blackmail/Extortion/Threat
1 03 Kidnapping
1 04 Manslaughter by negligence
1 05 Murder/Homicide (Non-negligent Manslaughter)
1 06 Robbery
1 07 Sexual Assault/Rape by force
1 08 Weapons
1 09 Domestic Violence
1 10 Child Abuse
1 11 Spouse/Partner Abuse
1 12 Child Neglect
1 13 Violation of Protection Order

DRUG/ALCOHOL RELATED OFFENSES	PROPERTY OFFENSES	MISCELLANEOUS OFFENSES
2 01 Driving While Intoxicated/ Driving Under the Influence	3 01 Arson	5 01 Commercial Sex/Prostitution
2 02 Drug Possession	3 02 Bribery	5 02 Embezzlement
2 03 Drug Sale	3 03 Burglary	5 03 Fare Beating
2 04 Liquor violation	3 04 Burglary Tools	5 04 Flight/Escape/Bench Warrant
2 05 Possession of alcohol	3 05 Damage/Destroy Property	5 05 Gambling
2 06 Under the influence of a Controlled Substance	3 06 Forgery	5 06 Obscenity (e g , indecent exposure)
	3 07 Fraud	5 07 Obstructing Police/Resisting Arrest
	3 08 Larceny/Theft	5 08 Other (specify above)
	3 09 Stolen Property	5 09 Public Peace/Disturbance/Mischief/ Reckless Endangerment
	3 10 Stolen Vehicle	5 10 Pickpocket/Jostling
	3 11 Trespassing	5 11 Other Sex Offenses
		5 12 Unspecified Probation/Parole/ROR Violation

NOTE: INTERVIEWER INSTRUCTIONS ARE IN CAPITAL LETTERS. READ ANSWER CHOICES TO THE RESPONDENT ONLY WHEN INSTRUCTED TO DO SO. EVERYTHING ELSE IN LOWER CASE OR INITIAL CAPS MUST BE READ TO RESPONDENT. USE PEN, NOT PENCIL TO COMPLETE INTERVIEW.

Interviewer's Initials: _____   INTERVIEW START TIME __ __ : __ __ [ ] AM [ ] PM

READ AS WRITTEN: This interview is part of a [x-agency] funded study. Your participation is voluntary. The information you provide is confidential and anonymous and it will not help or hurt your case. At the end of the interview I will ask you to provide a urine sample. Can we begin the interview now?	**(CIRCLE ONE)** 1 Agreed to interview 2 Declined 3 Not available (ill, asleep, taken to court) 4 Other reason not interviewed (Specify) _____
THE INTERVIEW SHOULD ONLY BE CONDUCTED IN THE LANGUAGE USED ON THE FORM. CIRCLE THE CORRECT NUMBER AT THE RIGHT:	**(CIRCLE ONE)** 1 Interview conducted in the language on this form 2. No form in appropriate language (specify language needed) _____ 3 No interviewer available to conduct the interview in the appropriate language 4 Other (specify)
1. What day and what time were you arrested?	__ __ / __ __ __ __ : __ __ AM PM DAY MONTH CALCULATE # OF HOURS. IF MORE THAN 48 HOURS AGO, TERMINATE INTERVIEW __ __ HOURS AGO
2. What is the highest grade in school you have <u>completed</u>?	__ __ **EXAMPLES: NEVER ATTENDED SCHOOL =00; 9TH GRADE=09; H.S./SECONDARY GRAD =12; 1 YEAR COLLEGE=13;**
3. Did you graduate from secondary school/high school or get a certificate equal to secondary school?	**(CIRCLE ONE)** 0 NO 1 YES, SECONDARY SCHOOL **(SKIP TO Q5)** 2 YES, RECEIVED A CERTIFICATE **(SKIP TO Q5)**
4. Are you currently in secondary school/high school, college or some other training or education program?	**(CIRCLE ONE)** 0 NO 1 YES
5. What is your current marital status?	**(READ ALL CATEGORIES, CIRCLE ONE)** 1 Single, and have <u>never</u> been married 2 Married 3 Separated or divorced 4 Widowed

6. In the last 30 days, where did you live most of the time?	(READ ALL CATEGORIES, CIRCLE ONE) 1 In your own house or apartment 2 In someone else's house or apartment 3 In a shelter or emergency housing **(SKIP TO Q9)** 4 In jail or prison? **(SKIP TO Q9)** 5 In a halfway house or honors facility? **(SKIP TO Q9)** 6 In a drug or alcohol treatment program? **(SKIP TO Q9)** 7 On the street or no fixed address? **(SKIP TO Q9)**
7. Do you live in subsidized housing? That is, where you or your family qualifies for reduced housing payments or financial assistance based on your income?	(CIRCLE ONE) 0 NO 1 YES 9 DON'T KNOW
8. In the last 30 days, how many people have lived in your household on a regular basis, including yourself? (1= SELF, IF GREATER THAN 1 ASK 8A) **A. IF ARRESTEE SAYS MORE THAN ONE, ASK—** How are they related to you?	TOTAL NUMBER OF PEOPLE IN HOUSEHOLD INCLUDING THE RESPONDENT: __ __ _____ 1 MOTHER/STEPMOTHER _____ 2 FATHER/STEPFATHER _____ 3 SPOUSE _____ 4 BOY/GIRLFRIEND _____ 5 BIOLOGICAL CHILDREN _____ 6 ADOPTED OR STEPCHILDREN _____ 7 OTHER RELATIVES _____ 8 CHILDREN OF BOYFRIEND/GIRLFRIEND _____ 9 FRIENDS/ROOMMATES/OTHER UNRELATED (CIRCLE ALL THAT ARE MENTIONED AND GIVE THE NUMBER OF EACH)
9. How many children have you [given birth to/fathered]?	____ NUMBER OF BIOLOGICAL CHILDREN
10. We are interested in knowing the kinds of things you did to get money in the last 30 days. **ASK FOR EACH INCOME TYPE, A - D** Did you receive money from (INCOME SOURCE) in the last 30 days? **CIRCLE YES OR NO FOR EACH TYPE** **A. IF ANY CATEGORY IS ANSWERED YES—** How much did you earn from all legal sources during the past 30 days [Include the value of government payments plus the net value of any job earnings.]	NO YES A. Welfare or other regular government payment? 0 1 B. Working full time the whole month? 0 1 C. Working part time or doing odd-jobs? 0 1 D. Other means? (SPECIFY) 0 1 _____ *(Total in local currency units)* _____
11. How much money did you get from your family or friends in the past 30 days?	*(Total in local currency units)* _____
12. We're also interested in knowing the kinds of illegal things you did to get money in the last 30 days. **ASK FOR EACH INCOME TYPE, A - D** Did you earn money from (INCOME SOURCE) in the past 30 days? **CIRCLE THE CODE FOR YES OR NO FOR EACH TYPE** **A. IF ANY CATEGORY IS ANSWERED YES—** How much were your earnings or profits from all illegal sources during the past 30 days?	*(Use local terms as appropriate)* NO YES A. Prostitution or hustling? 0 1 B. Dealing, growing or manufacturing drugs? 0 1 C. Doing other illegal things such as stealing, selling stolen property or running scams? 0 1 D. Other means? 0 1 Specify _____ *(Total in local currency units)* _____

13. DRUG GRID

(CIRCLE "0" FOR NO AND "1" FOR YES)

AGE_____ (CALCULATE FROM Y.O.B. AND VERIFY WITH ARRESTEE)	Alcohol	Tobacco	Marijuana	Crack	Powder Cocaine	Heroin, Black Tar, Dilaudid, Morphine (opiates)	PCP/Angel Dust	Amphetamines/Speed	Downers/Barbiturates	Quaaludes/Ludes	Methdone	Crystal Meth.	Valium or other Tranquilizers	LSD/Acid	Inhalents
11 Have you ever tried any of the following drugs? (READ ALL DRUGS)	0 1	0 1	0 1	0 1	0 1	0 1	0 1	0 1	0 1	0 1	0 1	0 1	0 1	0 1	0 1
PROCEED DOWN THE COLUMN FOR EACH DRUG THE ARRESTEE EVER TRIED															
12 When you first tried (NAME DRUG) how old were you?															
13 Have you used (NAME DRUG) during the last 12 months?	0 1	0 1	0 1	0 1	0 1	0 1	0 1	0 1	0 1	0 1	0 1	0 1	0 1	0 1	0 1
IF NO, SKIP TO QUESTION 18 (SKIP MAY DIFFER FOR EACH DRUG)															
14 In the past 3 days did you use (NAME DRUG)?	0 1	0 1	0 1	0 1	0 1	0 1	0 1	0 1	0 1	0 1	0 1	0 1	0 1	0 1	0 1
15 How many days did you use (NAME DRUG) in the past 30 days?															
16 During the past 12 months, have you consciously tried to cut down or quit using (NAME DRUG) on your own?	0 1	0 1	0 1	0 1	0 1	0 1	0 1	0 1	0 1	0 1	0 1	0 1	0 1	0 1	0 1
A If yes, were you successful?	0 1	0 1	0 1	0 1	0 1	0 1	0 1	0 1	0 1	0 1	0 1	0 1	0 1	0 1	0 1
17 Have you felt that you needed or were dependent on (NAME DRUG) in the past 12 months?	0 1	0 1	0 1	0 1	0 1	0 1	0 1	0 1	0 1	0 1	0 1	0 1	0 1	0 1	0 1
18 Are you now receiving treatment or detox for (NAME DRUG)?	0 1	0 1	0 1	0 1	0 1	0 1	0 1	0 1	0 1	0 1	0 1	0 1	0 1	0 1	0 1
19 Have you received treatment or detox for (NAME DRUG) in the past?	0 1	0 1	0 1	0 1	0 1	0 1	0 1	0 1	0 1	0 1	0 1	0 1	0 1	0 1	0 1
20 Do you feel you could use treatment for (NAME DRUG)?	0 1	0 1	0 1	0 1	0 1	0 1	0 1	0 1	0 1	0 1	0 1	0 1	0 1	0 1	0 1

14. Are there any <u>other</u> drugs that you have used illegally in the past 30 days?

(CIRCLE ONE)
0 NO
1 YES (Specify) _____

15. Now I am going to ask about your experience with drug and alcohol treatment.

A. Have you ever participated in a [TREATMENT TYPE] program? *(Use locally appropriate treatment types)*	(IF YES) B. How many times have you entered a [TYPE] program?	C. Are you currently in [TYPE]? (IF NO) What year were you most recently in [TYPE]?	D. What type of substance abuse (including alcohol) [are/were] you mainly being treated for [the most recent time you were in treatment]?
TYPE	A EVER	B # OF TIMES	C CURRENTLY/YR	D TYPE OF SUBSTANCE
	NO	YES		YES	YEAR	
Detox						
Residential						
Outpatient						
Methadone Maint.						
Self Help (AA, CA)						
Spiritual or Religious						

16.	Had you been drinking alcohol when you got involved in the activities for which you were just arrested?	(CIRCLE ONE) 0 NO 1 YES
17.	Had you recently used any illegal drugs when you got involved in the activities for which you were just arrested? A. What drug(s) had you used?	CIRCLE ONE) 0 NO **(SKIP TO Q. 18)** 1 YES **(CIRCLE ALL THAT APPLY)** 1. Marijuana 2. Cocaine/Crack/Cocaine paste 3. Heroin/opiates 4. Amphetamines 5. Other (Specify) _____
18.	Had you been looking for, or trying to buy or sell, any illegal drugs when you got involved in the activities for which you were just arrested? A. What drug(s) had you used?	CIRCLE ONE 0 NO **(SKIP TO Q. 19)** 1 YES **(CIRCLE ALL THAT APPLY)** 1. Marijuana 2. Cocaine/Crack/Cocaine paste 3. Heroin/opiates 4. Amphetamines 5. Other (Specify) _____
19.	Have you ever made money or obtained drugs by selling illegal drugs or being involved in the manufacture or transportation of drugs?	(CIRCLE ONE) 0 NO 1 YES

The next several questions are about previous arrests, not including the arrest for which you are here now.

20.	During the past 12 months, have you been arrested and booked for breaking a law, whether or not you were guilty?	(CIRCLE ONE) 0 NO **(SKIP TO Q. 23)** 1 YES
21.	How many times have you been arrested in the past 12 months?	___ ___ NUMBER OF ARRESTS
22.	What were the arrest charges?	
23.	In the past 12 months, have you served time in jail or prison on a sentence?	(CIRCLE ONE) 0 NO 1 YES
24.	How would you describe your racial or ethnic background? Please answer "yes" or "no" to each of the following. You can answer "yes" to more than one. Do you consider yourself to be?	*(Use racial/ethnic groupings appropriate to local situation)* 1. 2. 3. 4. 5. 6. 7. 8.

INTERVIEW COMPLETION TIME ___ ___ : ___ ___ [ ] AM [ ] PM

Now I need for you to take this specimen container to [DESIGNATED AREA] and provide a urine specimen.

SPECIMEN (AFTER ANY ADDITIONAL QUESTIONS/ADDENDA) WAS:

0 Refused/did not try   1 Provided   2 Tried/could not produce specimen

Appendix B

Demographic, Economic, and Drug Information by Country

Exhibit B–1. General Demographic Information for I–ADAM Countries

	Australia	Chile	England and Wales	Malaysia
Geographic area	7,692,030 km² (3,076,812 mi²)	756,950 km² (302,780 mi²)	145,915 km² (58,366 mi²)	330,113 km² (132,045 mi²)
Political subdivisions	6 States and 2 Territories	13 regions (Regions I–XII and Santiago Municipal Region)	England: 39 counties, 7 metropolitan counties; Wales: 8 counties	13 States and 2 Federal Territories
Estimated population	19.1 million	15,153,797	51.1 million	23.3 million
Population by racial or ethnic origin	92% white, 5.9% Asian, 2.1% aboriginal/indigenous (24% born outside Australia)	95% white/white-Amerindian, 3% Amerindian, 2% other	94.0% white, 2.9% South Asian, 1.8% black, 0.7% East Asian, 06% other	58.1% Malay, 24.3% Chinese, 6.9% Indian, 3.2% other, 7.5% noncitizens
Religious affiliation	50.4% Protestant, 26% Roman Catholic, 1% non-Christian, 12.6% unaffiliated	89% Roman Catholic, 11% Protestant, <1% Jewish	48% Protestant, 15% Roman Catholic, 1.7 % Muslim, 0.7% Hindu, 0.5% Sikh, 35% unaffiliated (U.K.)	Islam (official religion), Buddhism, Daoism, Hinduism, Christianity, and Sikhism
Gender distribution	50.2% female, 49.8% male	50.5% female, 49.5% male	51.1% female, 48.9% male	51.2% female, 48.8% male
Age distribution (years)	0–14, 20.6%; 15–64, 67.1%; ≥65, 12.3%	0–14, 28%; 15–64, 65%; ≥65, 7%	0–15, 20.3%; 16–64, male/ 16–59, female, 61.5%; ≥65 male/ ≥60 years, 18.1%	0–19, 43.2%; 20–39, 33.0%; 40–59, 17.7%; ≥60, 6.1%
Life expectancy (years)	78.6, average; 75.2, male; 81.9, female	75.74, average; 72.43, male; 79.22, female		72.2, average; 70.2, male; 75.0, female
Population growth rate	1.1%	1.2%	0.3% (U.K.)	2.3%
Birth rate	13.08 per 1,000	17.19 per 1,000	13.7 per 1,000	24.5 per 1,000
Total fertility rate (per woman)	1.8 children	2.2 children	1.7 children	3.3 children
Death rate	5.9 per 1,000	5.5 per 1,000	11.2 per 1,000	5.3 per 1,000
Literacy rate (age 15 and older) (1999)	>99%	95.2% total, 95.4% male, 95.0% female	>99%	86.4% total, 90.7% male, 82.0% female
Languages	English	Spanish	English, Welsh (26% of of population of Wales)	Bahasa Malaysia English, Chinese dialects (principally Cantonese and Mandarin), Tamil, Telagu, Malayalam, Panjabi, Thai, and indigenous languages
Urbanization rate	86%	85.4%	89% (U.K.)	57%
Human development index (1999)	0.936 (world rank: 2)	0.825 (world rank: 39)	0.923 (world rank: 14)*	0.774 (world rank: 8)

*The human development index given for England and Wales and for Scotland is for the United Kingdom as a whole.

Exhibit B–1. General Demographic Information for I–ADAM Countries (continued)

	The Netherlands	Scotland	South Africa	United States
Geographic area	41,526 km² (16,610 mi²)	78,789 km² (31,510 mi²)	1,219,000 km² (487,636 mi²)	9,629,047 km² (3,717,796 mi²)
Political subdivisions	12 provinces	9 regions and 3 islands areas	9 provinces	50 States and District of Columbia
Estimated population	15.7 million	5.1 million	43.1 million	281.4 million
Population by racial or ethnic origin	91% Dutch, 9% Moroccan Turks, other	98.7% white, 0.7% South Asian, 3% East Asian, 0.1% black, 0.2% other	77% African, 11% white, 9% colored (mixed white and African or Malay), 3% Indian	75.1% white, 12.3% black 3.6% Asian, 0.9% American Indian and Alaska Native 0.1% Pacific Islander, 5.5% other, 2.4% multiracial (12% Hispanic—may be categorized as any race)
Religious affiliation	34% Roman Catholic, 25% Protestant, 3% Muslim, 36% unaffiliated	48% Protestant, 15% Roman Catholic, 1.7 % Muslim, 0.7% Hindu, 0.5% Sikh, 35% unaffiliated (U.K.)	68% Christian, 28.5% indigenous beliefs and animist, 2% Muslim 1.5% Hindu	55% Protestant, 28% Roman Catholic, 2% Jewish, 6% other, 8% unaffiliated
Gender distribution	51% female, 49% male	50.8% female, 49.5% male (U.K.)	52% female, 48% male	50.9% female, 49.1% male
Age distribution (years)	0–14, 18%; 15–64, 68%; ≥65, 14%	0–15, 19.9%; 15–64, 64.7%; ≥65, 15.4%	0–19, 47%; 20–39, 35%; 40–69, 16%; ≥70, 2%	0–14, 21.4%; 15–64, 66.1%; ≥65, 12.4%
Life expectancy (years)	78.28, average; 75.4, male; 81.28, female	75.1, average; 72.1, male; 78.1, female	57–64, average	77.1, average; 74.2, male; 79.9, female
Population growth rate	0.690%	−0.1% (Scotland)	2.1%	0.990%
Birth rate	12.1 per 1,000	10.6 per 1,000	24.6 per 1,000	14.2 per 1,000
Total fertility rate (per woman)	1.6 children	1.5 children	2.5 children	2.1 children
Death rate	8.72 per 1,000	11.67 per 1,000	11.89 per 1,000	8.8 per 1,000
Literacy rate (age 15 and older) (1999)	99%	>99%	81%	>99%
Languages	Dutch	English, Scottish Gaelic (60,000 speakers)	IsiZulu, 24%; IsiXhosa, 17.5%; Afrikaans, 15.1%; Sepedi, 9.8%; English, 9.1%; Setswana; Sesotho; Xitsonga; siSwazi; IsiNdebele; Tshivenda	English
Urbanization rate	89%	89% (U.K.)	54%	89%
Human development index (1999)	0.931 (world rank: 8)	0.923 (world rank: 14)*	0.702 (world rank: 94)	0.936 (world rank: 6)

*The human development index given for England and Wales and for Scotland is for the United Kingdom as a whole.

Exhibit B-2. General Economic Information for I-ADAM Countries

	Australia	Chile	England and Wales*	Malaysia
Gross domestic product (GDP) (annual)	$416.2 billion	$185.1 billion	$1.29 trillion	$171 billion
GDP growth rate	1.6%	–1%	1.9%	5.8%
GDP per capita	$22,200	$12,400	$21,800	$3,238
GDP composition by sector	Services, 71%; industry, 26%; agriculture, 3%	Services, 61%; industry, 33%; agriculture, 6%	Services, 73%; industry, 25.3%; agriculture, 1.7%	Industry, 46%; services, 42%; agriculture, 12%
Major export partners	Japan, 20%; European Union, 14%; Association of Southeast Asian Nations (ASEAN), 11%; United States, 10%	European Union, 27%; United States, 16%; Japan, 14%; Brazil, 6%; Argentina, 5%	European Union, 58% (Germany, 12%; France, 10%; Netherlands, 8%); United States, 13%	United States, 23%; Singapore, 16%; Japan, 11%; Hong Kong, 5%; Netherlands, 5%; Taiwan, 5%; Thailand, 3%
Major export commodities	Coal, gold, meat, wool, alumina, iron ore, wheat, machinery, and transport equipment	Copper, fish, fruits, paper, pulp, and chemicals	Manufactured goods, fuels, chemicals, food, beverages, and tobacco	Electronic equipment, petroleum and liquefied natural gas, chemicals, wood and wood products, rubber, textiles
Labor force	8.9 million	5.8 million	29.2 million	9.3 million
Labor force composition by sector	Services, 73%; industry, 22%; agriculture, 5%	Services, 59%; industry, 27%; agriculture, 14%	Services, 69%; manufacturing and construction, 18%	Manufacturing, 27%; agriculture, forestry, and fisheries, 16%; local trade and tourism, 17%; services, 15%; government, 10%; construction, 9%
Poverty rate	13.0%	22%	18%	15.5%
Unemployment rate	6.8%	9%	8.4%	3.3%

*Figures provided for England and Wales and for Scotland relate to the United Kingdom as a whole.

Exhibit B–2. General Economic Information for I–ADAM Countries (continued)

	The Netherlands	Scotland	South Africa	United States
Gross domestic product (GDP) (annual)	$365.1 billion	$1.29 trillion	$135 billion	$9.256 trillion
GDP growth rate	3.4%	1.9%	1.2%	4.1%
GDP per capita	$23,100	$21,800	$3,215	$33,833
GDP composition by sector	Services, 69.7%; industry, 26.8%; agriculture, 3.5%	Services, 73%; industry, 25.3%; agriculture, 1.7%	Services, 60%; industry, 35%; agriculture, 5%	Industry, 80%; services, 18%; agriculture, 2%
Major export partners	Germany, 27%; Belgium-Luxembourg, 13%; France, 11%; United Kingdom, 10%; Italy, 6%; Central and Eastern Europe; United States	European Union, 58%; Germany, 12%; France, 10%; Netherlands, 8%; United States, 13%	United Kingdom, Italy, Japan, United States, Germany	Canada, 23%; Mexico, 12%; Japan, 8%; United Kingdom, 6%; Germany, 4%; France, 3%; Netherlands, 3%
Major export commodities	Machinery and transport equipment, chemicals and fuels, foodstuffs, clothing	Manufactured goods, fuels, chemicals, food, beverages, and tobacco	Gold, diamonds, other metals and minerals, machinery and equipment	Capital goods, automobiles, industrial supplies and raw materials, consumer goods, agricultural products
Labor force	7 million	29.2 million	15 million	139.4 million
Labor force composition by sector	Services, 73%; industry, 23%; agriculture, 4%	Services, 68.9%; manufacturing and construction, 17.5%; government, 11.3%; energy, 1.2%; agriculture, 1.1%	Services, 45%; agriculture, 30%; industry, 25%	Managerial/professional, 30.3%; technical, sales, and administrative support, 29.2; services, 15%; government, 10%; services 13.4%; manufacturing, mining, transportation, and crafts 24.5%
Poverty rate	10.5%	18%	28.5% (based on imputed expenditures per household < R800 per month)	11.8%
Unemployment rate	3.3%	8.4%	34%	4.5%

*Figures provided for England and Wales and for Scotland relate to the United Kingdom as a whole.

Exhibit B–3. Prevalence (Percentage) of Drug Use in the General Adult Population in I–ADAM Countries

	Australia		Chile		England and Wales		The Netherlands		Scotland		United States	
	Lifetime use	Past 12-month use	Lifetime use	Past 12-month use	Lifetime use	Past 12-month use	Lifetime use	Past 12-month use	Lifetime use	Past 12-month use	Lifetime use	Past 12-month use
Any drugs	48 (male) 41 (female)	24 (male) 19 (female)	17.45	5.31	32	11	—	—	22.5	9.0	39.7	11.9
Cannabis (marijuana)	39	18	16.77	4.73	25	9	15.6	4.5	19.0	8.2	39.7	11.9
Amphetamines	9	4			10	3	1.9	0.3	7.3	3.1	7.2	1.0
Powder cocaine	4	1	4.02	1.32	3	1	2.1	0.6	2.6	1.0	11.5	1.7
Crack cocaine	—	—	2.27	0.81	1	<1	—	—	0.7	0.2	2.7	0.5
LSD	10	3	—	—	5	1	1.8	0.3	5.5	1.3	8.7	1.4
Other hallucinogens	10	3	—	—	5	1	1.6	0.6	5.1	1.3	11.3	1.4
Ecstasy (MDMA)	5	2	—	—	4	1	1.9	0.7	4.0	2.1	4.3	0.5
Barbiturates	2	—	12.52	6.89	3	1	19.6	8.2	3.7	1.1	9.8	1.6
Heroin	2	1			1	<1	0.3	0.1	0.8	0.1	1.4	0.2
Methadone	—	—	—	—	<1	<1	—	—	0.9	0.3	—	—
Solvents/inhalants	—	—	—	—	2	<1	0.5	0.1	2.1	0.2	7.8	1.1
Other	—	—	—	—	—	—	7.3	2.6	4.2	0.9	11.6	3.1

Notes: Figures for Malaysia are not available. For information on drug use in South Africa, see "The Nature and Extent of Drug Use and Related Problems in South Africa: SARPA National Surveillance Project," available online at http://www.nrf.ac.za/netpro/docs/lee_sarpa_project_1997.doc.

— = Not available.

About the National Institute of Justice

NIJ is the research and development agency of the U.S. Department of Justice and is the only Federal agency solely dedicated to researching crime control and justice issues. NIJ provides objective, independent, nonpartisan, evidence-based knowledge and tools to meet the challenges of crime and justice, particularly at the State and local levels. NIJ's principal authorities are derived from the Omnibus Crime Control and Safe Streets Act of 1968, as amended (42 U.S.C. §§ 3721–3722).

NIJ's Mission

In partnership with others, NIJ's mission is to prevent and reduce crime, improve law enforcement and the administration of justice, and promote public safety. By applying the disciplines of the social and physical sciences, NIJ—

- **Researches** the nature and impact of crime and delinquency.
- **Develops** applied technologies, standards, and tools for criminal justice practitioners.
- **Evaluates** existing programs and responses to crime.
- **Tests** innovative concepts and program models in the field.
- **Assists** policymakers, program partners, and justice agencies.
- **Disseminates** knowledge to many audiences.

NIJ's Strategic Direction and Program Areas

NIJ is committed to five challenges as part of its strategic plan: 1) **rethinking justice** and the processes that create just communities; 2) **understanding the nexus** between social conditions and crime; 3) **breaking the cycle** of crime by testing research-based interventions; 4) **creating the tools** and technologies that meet the needs of practitioners; and 5) **expanding horizons** through interdisciplinary and international perspectives. In addressing these strategic challenges, the Institute is involved in the following program areas: crime control and prevention, drugs and crime, justice systems and offender behavior, violence and victimization, communications and information technologies, critical incident response, investigative and forensic sciences (including DNA), less-than-lethal technologies, officer protection, education and training technologies, testing and standards, technology assistance to law enforcement and corrections agencies, field testing of promising programs, and international crime control. NIJ communicates its findings through conferences and print and electronic media.

NIJ's Structure

The NIJ Director is appointed by the President and confirmed by the Senate. The NIJ Director establishes the Institute's objectives, guided by the priorities of the Office of Justice Programs, the U.S. Department of Justice, and the needs of the field. NIJ actively solicits the views of criminal justice and other professionals and researchers to inform its search for the knowledge and tools to guide policy and practice.

NIJ has three operating units. The Office of Research and Evaluation manages social science research and evaluation and crime mapping research. The Office of Science and Technology manages technology research and development, standards development, and technology assistance to State and local law enforcement and corrections agencies. The Office of Development and Communications manages field tests of model programs, international research, and knowledge dissemination programs. NIJ is a component of the Office of Justice Programs, which also includes the Bureau of Justice Assistance, the Bureau of Justice Statistics, the Office of Juvenile Justice and Delinquency Prevention, and the Office for Victims of Crime.

To find out more about the National Institute of Justice, please contact:

National Criminal Justice Reference Service
P.O. Box 6000
Rockville, MD 20849–6000
800–851–3420
e-mail: *askncjrs@ncjrs.org*

To obtain an electronic version of this document, access the NIJ Web site
(*http://www.ojp.usdoj.gov/nij*).

If you have questions, call or e-mail NCJRS.

U.S. Department of Justice
Office of Justice Programs
National Institute of Justice

Washington, DC 20531

U.S. Department of Justice
Office of Justice Programs
National Institute of Justice

www.ingramcontent.com/pod-product-compliance
Lightning Source LLC
Chambersburg PA
CBHW081118170526
45165CB00008B/2481